A BEGINNER'S GUIDE TO PRODUCING TV

A BEGINNER'S GUIDE TO PRODUCING TV

Complete Planning Techniques and Scripts to Shoot

James R. Caruso

Mavis E. Arthur

PRENTICE HALL, Englewood Cliffs, New Jersey 07632

Library of Congress Cataloging-in-Publication Data

CARUSO, JAMES R., (date)
    A beginner's guide to producing TV.

    Includes index.
    1. Television—Production and direction.   2. Video recordings—Production and direction
I. Arthur, Mavis E., (date).   II. Title.
PN1992.75.C33 1990   791.45′0232   88-28877
ISBN   0-13-944091-7
ISBN   0-13-944109-3 (pbk.)

Cover design: Lundgren Graphics, Ltd.
Manufacturing buyer: Bob Anderson

The publisher offers discounts on this book when ordered in bulk quantities.
For more information, write:

Special Sales/College Marketing
Prentice Hall
College Technical and Reference Division
Englewood Cliffs, New Jersey 07632

Printed in the United States of America

10   9   8   7   6   5   4   3   2   1

ISBN 0-13-944091-7
ISBN 0-13-944109-3 {PBK}

Prentice-Hall International (UK) Limited, *London*
Prentice-Hall of Australia Pty. Limited, *Sydney*
Prentice-Hall Canada Inc., *Toronto*
Prentice-Hall Hispanoamericana, S.A., *Mexico*
Prentice-Hall of India Private Limited, *New Delhi*
Prentice-Hall of Japan, Inc., *Tokyo*
Simon & Schuster Asia Pte. Ltd., *Singapore*
Editora Prentice-Hall do Brasil, Ltda., *Rio de Janeiro*

CONTENTS

CHAPTER FOURTEEN
101 IDEAS FOR VIDEOS

CHAPTER FIFTEEN
SCRIPTS TO SHOOT

APPENDIX

INDEX

PREFACE

Congratulations! You are about to become a producer of television shows rather than a warehouser for hours and hours of used videotape. You have made a smart decision to learn how to make the most of your equipment. By the time you finish this book, you will be able to produce "real" TV that is entertaining and that will, no doubt, get you rave reviews. Rather than taping everything in sight, ending up with hours and hours of videotape that no human being will probably ever see more than once—no one could live long enough to watch it twice—you will learn to put a little thought and planning into your video taping and create your own professional-looking videos. From opening titles to end credits, you will be a pro. It is possible and it *is* easy!

The secret of a good video is planning and follow-through. Anyone can create a good video with organization. Think of it as a story you want to tell, and then think about how you want to tell it. It is as simple as that: thinking about what you want to do *before* you do it.

Making a good video is as simple as answering some very basic questions:

1. What do you want to do?
2. What equipment is available to you?
3. What other things can you add to make the show more exciting?
4. How do you put your ideas together to make a show?
5. Are you ready to roll tape?

You will be able to answer these questions by using the easy steps to planning in this book. You will also learn techniques of formatting and scripting your own shows. And then, to get you off to a good start, we have included some scripts you can shoot. Choose a script, make your plans using the checklists, and roll tape. See for yourself what a difference organization can make. Then use the other scripts included here or write your own. And. . .

Welcome to show biz! You are about to become a star or at least create a few.

INTRODUCTION

HOW TO USE THIS BOOK

This book contains: (1) *complete* planning techniques for making a better video, including ideas for videos and (2) scripts to shoot. It is designed so that you can pick and choose the aspects of video production in which you are interested in improving your skills. You may elect to read and follow absolutely every step of the planning techniques or you may elect to skip portions entirely and go directly to one particular step that is of interest to you.

We do suggest that you read Chapter 1—the crash course in TV talk—and that you check out the scripts included at the back of this book. They are there just for you to produce in your own style. You are, after all, the Producer *and* Director.

A BEGINNER'S GUIDE TO PRODUCING TV

CHAPTER ONE

LEARNING TO TALK TV

Creating a TV show means more than just rolling tape. A good TV show is the successful combination of many different pieces, as you can see in Figure 1–1. You will not necessarily need to do absolutely every one of these to make your video. You will discover, however, that the more you do the more you will understand just what it takes to make a good video, and just how important good planning is to the end product. We will cover all aspects of planning in this book. You will be the final judge when it comes to choosing which planning steps to use. Once you are aware of all the steps that might make the production easier and better, you will be able to make a more informed decision.

Before you get started on your first professional production, let's take a crash course in TV Talk. There are a million and one words coined just for television. TV is, after all, a creative business and is, therefore, generating new words all the time. Luckily, however, you don't have to know them all. The following crash course will allow you to talk TV and, at the same time, better understand how television is made.

WHAT ARE THE WORDS USED IN TV?

First of all, when you **AIR A SHOW,** you show it on a television set to the intended audience. Any show that is going to air on commercial television to a national audience is said to be **BROADCAST.** Any show that is produced for small, specialized audiences with limited viewership, whether for a company convention or privately just for you in your own living room, is **NON BROADCAST.** A show is said to be **BROADCAST QUALITY** when it is good enough technically and creatively to be aired on broadcast television. We will not deal with the technical or creative requirements for a broadcast production in this book. We mention it here only because many of the tape formats and equipment available today to the nonbroadcast market are equivalent to those available to the broadcast market. It is possible, therefore, for anyone to make a broadcast-quality production with the right planning and execution.

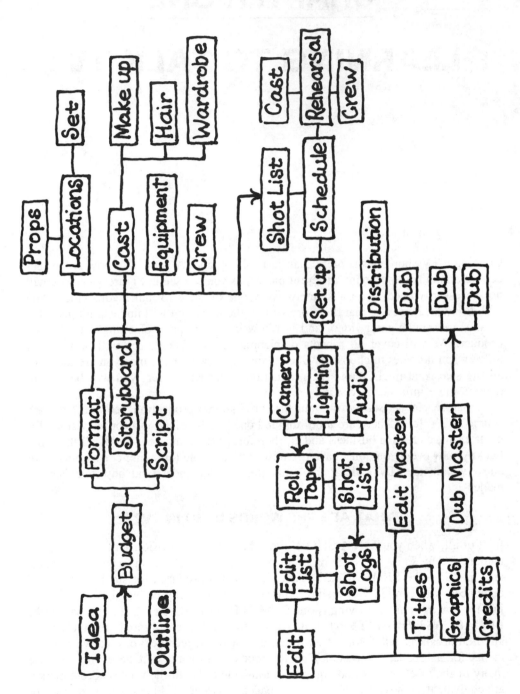

Figure 1–1   The planning steps to creating a TV show.

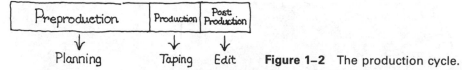

Figure 1–2 The production cycle.

From the moment you conceive the idea for the show until the time you actually begin taping, you are in **PRE PRODUCTION.** When you are taping, you are in **PRODUCTION.** After you have everything recorded on tape and are in the process of putting it all together to make the show, you are in **POST PRODUCTION** (Figure 1–2).

TV PERSPECTIVE means seeing the picture as it will be seen on a television set. A television set is not a square, but more of a rectangle with an **ASPECT RATIO** of three high to four wide, as shown in Figure 1–3. It thus shows more of the horizontal than the vertical. A square two inches by two inches shown on a three-inch by four-inch television screen would leave a one-inch border on each side but only a half inch border on the top and bottom. This TV perspective must be part of your thinking when planning shots.

VIDEOTAPE, sometimes just called **TAPE,** is a thin piece of plastic covered with a powder that is magnetized as the electronic impulses of pictures and sound are recorded. Sound and pictures are recorded on different **TRACKS** on the tape. There is always one track for video, but there may be one or two tracks for audio. With two audio tracks, you have the ability to record two different audio elements and later either mix these two together to make one soundtrack on the playback or play one or the other of the tracks.

There are different sizes of videotape, as shown in Figure 1–4. The size is measured by the actual width of the tape. On the standard VCR, the tape required may be either **VHS** or **BETA** or **8MM** format. One tape format will not play on another format machine, i.e. VHS will not play on a BETA machine and vice versa. VHS or BETA tape is also sometimes referred to as **1/2″ TAPE;** 1/2″ is the width of the tape. **8MM** tape is **1/4″,** and in the broadcast industry, there is also **3/4″, 1″,** and **2″** tape. 2″ is sometimes called **QUAD.**

Pictures and sound can be recorded on standard VCRs at different speeds. This is simply a mechanical function of the machine.

Videotape may have or develop problems through the manufacturing or recording process that cause it to record improperly or poorly, as seen in Figure 1–5. For example, any tape may have **DROPOUT,** a loss of a part of the picture. Dropout

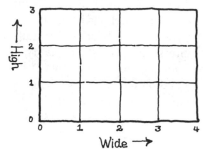

Figure 1–3 The aspect ratio of the television screen is 4 wide to 3 high.

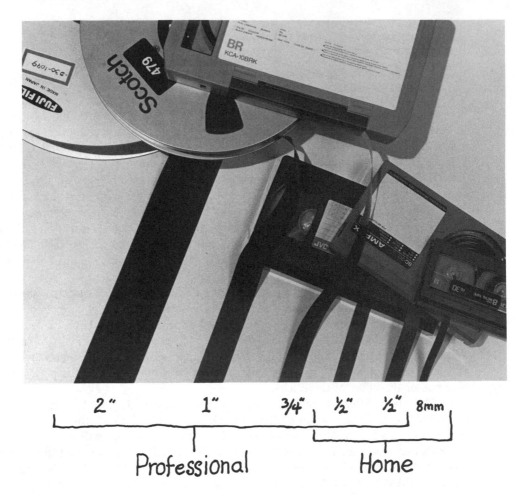

Figure 1–4 The sizes of videotape now available for both broadcast and nonbroadcast.

occurs when the powder surface on the tape is not distributed evenly, leaving "holes." The tape will not record anything where these holes exist. This dropout will show up on your video as white snowy spots or lines across the picture. Even the highest quality tape is likely to have some dropout acquired in the manufacturing process or it could develop the problem through continued playback or poor storage or bad recording techniques. There is nothing you can do about poor manufacturing, so it is best to stick to well known manufacturers that keep their quality standards high. Damage caused after you take the tape home can be avoided by (1) storing the tape upright, not on its side where objects stacked on top can crush the cassette, (2) rewinding the tape after every use so the tape is not exposed to damage at the recording position, and (3) repacking the tape by running fast-forward to the end and then rewinding to the front to prevent stretching.

NOISE Caused By Low Light Level.

DROPOUT Caused By Lack of Magnetic Particles on Tape.

Figure 1–5   Examples of tape problems.

Poor picture or sound quality may also be caused by a **GLITCH** or **HIT** or **NOISE,** all of which could be the result of clogged recording heads on your VCR, loose connecting cables, power surges, equipment malfunctions, or poor lighting. All create some distortion in the picture, either tearing it or losing it completely. A glitch is usually caused by poor tape or a problem with the recording equipment. A hit is caused by an electrical surge affecting the recording process. Noise is caused when the system is not getting enough information to record the scene and thus it essentially records its own electronic signal. Picture quality can also acquire more noise as you make repeated **DUBS** of the show. A dub is a copy of the show recorded on another piece of tape. Noise may show up as a fuzziness in the picture that is losing its line definition. This particular kind of noise is called **BLEEDING.** Bleeding becomes progressively worse as you make a dub of a dub of a dub of a dub, and so on.

The **ORIGINAL MATERIAL** is what is on the tape that was inside the camcorder when you recorded. Original material is sometimes called **RAW FOOTAGE,** meaning it has not been edited yet but is just as it was shot. The **EDIT MASTER** is the tape on which you edit the show. The **DUB MASTER** is a copy of the edit master and is used only to make the **DISTRIBUTION DUBS,** those copies of the show that will be viewed. This dub process saves the edit master from possibly being destroyed or damaged during duplication. Original material and edit masters are usually stored in a **VAULT**—a cool dry place—and pulled out only to create new dub masters when needed. A vault could be a garage or a closet where tape can be kept safely and at a relatively even temperature.

The video goes through certain **GENERATIONS** as it is recorded, edited, and dubbed. A generation is an indication of whether the recorded material is the original material or a copy, or a copy of a copy and so on, as indicated in Figure 1–6. The original material is **FIRST GENERATION,** meaning it is the first tape

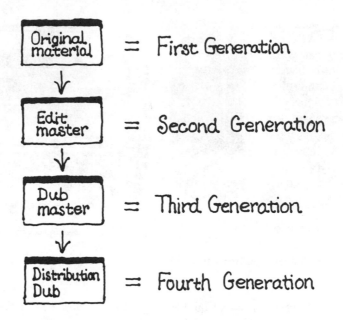

Figure 1-6   Tape generation: the technical quality of the show deteriorates with every step down.

on which the video has been recorded. A copy made from this first-generation material becomes **SECOND GENERATION.** A copy made from that second-generation material is **THIRD GENERATION.** A copy made from that third-generation copy is **FOURTH GENERATION** and so on. Edit masters are usually second generation, that is, a copy of the original first-generation material. Dub masters, being copies of the edit master, are then third generation and dubs from the dub master are fourth generation. You do not want to go beyond fourth generation if you can prevent it, as the picture quality will be deteriorating rapidly by then. As a matter of fact, if you are only making a few dubs, you might want to save a generation and make your distribution dubs from the edit master to save some of the picture quality you would lose by going down that extra generation to a dub master.

Sometimes you will want to **ROLL OVER** or **DUB** (duplicate) certain pieces of video to another tape for editing purposes. When you do this, you are taking that particular piece of video down one more generation. Your edit master may end up being a combination of second-generation and third-generation material if you use this roll-over technique.

When you are **ROLLING TAPE,** you are recording. In the professional television world, this is sometimes called **BURNING TAPE,** or using up the tape by recording on it. Burning can also mean erasing something you have already recorded. If you **BURN A TAKE,** you erase that shot. Since **TAPE IS CHEAP** compared to what it would cost to set up and reshoot, you don't want to erase shots unless you are in dire need of more tape on that particular cassette. When you **STOP TAPE** or **CUT,** you cease to record.

What you have already recorded is **FOOTAGE. STOCK FOOTAGE** is video you have stored for use as generic material for your future videos. It can be virtually anything from sunsets to family gatherings to people laughing.

If your VCR is **SHUTTLING,** it is moving fast-forward or in reverse to get to the place you want to be on the tape. If it is **CUEING UP,** it is shuttling to the position you have asked it to go to. Cueing up is usually done electronically by computer function and is not available on most consumer VCRs.

Anything you see on the show is the **VIDEO** and anything you hear is the **AUDIO,** as demonstrated simply in Figure 1–7. A **SPECIAL EFFECT** is something you create that is out of the ordinary. It may be a video EFX or an audio EFX, EFX being the abbreviation for special effect.

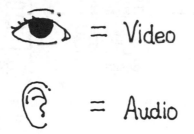

Figure 1–7   Video and audio—what you see and hear on a TV show.

The show is divided into **SCENES,** a scene being the action in one particular location. Within every scene there are different shots; as many as it takes to complete the scene. A **SHOT** is a sequence of events beginning when you roll tape and ending when you stop tape. A scene may move from one area to another and contain many different shots, but it generally becomes a new scene if the location changes. All spoken words are the **DIALOGUE.**

An **INTERVIEW** on camera is just that, one person interviewing another. A reporter talking to a city official for a piece on the evening news, for example. Interviews by their nature cannot be scripted but certainly the questions and subject of the interview can be planned. When the talent **IMPROVISES,** they are making it up as they go. This doesn't happen much in television but could on a breaking story or an impromptu interview situation.

When you are creating the idea for the show, you are writing an **OUTLINE.** When you are developing this outline into more specific plans, creating the show style, you are writing the **FORMAT.** The format is sometimes called the **STORY TREATMENT,** but in this book we will always refer to it as the format. When you write specific words for someone to say or plan specific shots you want to get, you are writing the **SCRIPT.**

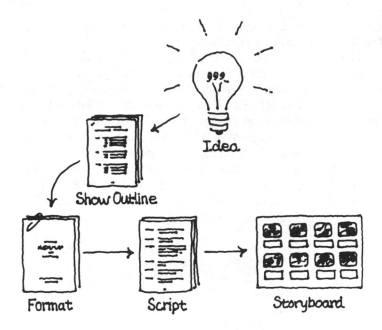

Figure 1–8   Idea to storyboard—making an idea into a show.

Figure 1–9   Storyboards—a pictorial representation of the show, usually hand drawn.

When you draw pictures to help visualize how the show will look when it is shot, you are **STORYBOARDING.** This development process from idea to storyboard is illustrated in Figure 1–8 and sample storyboards are shown in Figure 1–9.

When you are deciding who is going to star in the video, you are **CASTING** your **TALENT, STARS,** or **PERFORMERS.** All the performers have specific **ROLES;** that is, parts they are to play in the production. These roles fall into four basic categories, as seen in Figure 1–10. The main performers—the ones with the most dialogue—are your **PRINCIPAL TALENTS.** If you cast someone to do nothing but to say words and *never* appear on camera talking, that talent is your **VOICE-OVER ANNOUNCER.** An announcer that appears on camera speaking is an **ON-CAMERA ANNOUNCER.** Anyone who has a small speaking part is a

Principals
Major speaking/acting roles

Supporting Principals
Secondary speaking/acting roles

Announcer
On or off camera

Extras
Background

Figure 1–10   Casting the roles in the show.

SUPPORTING PRINCIPAL. All those who have nothing at all to say except as part of a group or crowd, principally acting as background, are your **EXTRAS.**

The clothing that your talent will wear is their **WARDROBE.** They may also wear **MAKEUP,** both men and women. Makeup can be as simple as face makeup or as elaborate as full-body makeup to create a monster. Any item that can be easily carried that is necessary to the action, such as a cigarette lighter to light a cigarette, is a **PROP.** The **SET** for the show is any large item you have built or acquired to do the show, from a special wall to a door, a table, or an entire house. All these elements are demonstrated in Figure 1–11.

The most important person in the production, as seen in Figure 1–12, is the person that puts up the money or raises the money to make the show, the **EXECUTIVE PRODUCER.** Reporting to the executive producer is the person actually ini-

Wardrobe
Clothing, jewelery, etc,
anything worn.

Make up
Face or full body,
Normal or exaggerated.

Prop
Easily carried items
necessary to action.

Set
Furniture, walls, etc.
built for show.

Figure 1–11   Wardrobe, makeup, props, and set for your show.

tiating the ideas and bringing all the elements together to make the show, the **PRO-DUCER.** The executive producer and the producer may be the same person. The person interpreting the script and actually deciding where and how the shots will be taped, who the performers will be, and who the crew will be is the **DIRECTOR.** Someone taking notes on the shoot or assisting the producer, the director, or a camera operator is a **PRODUCTION ASSISTANT.** The person preparing the format or script is the **WRITER.** Anyone running a camera is a **CAMERA OPERATOR.** A person planning the lighting is called the **LIGHTING DIRECTOR** or **L.D.** The lighting director's assistant, who does most of the physical labor installing and adjusting the lighting equipment, is called a **GAFFER.**

The **AUDIO DIRECTOR,** or **AD,** is the one who designs the audio requirements, including how many microphones will be needed, what kind they will be, and where they will be positioned, as well as whether any audio EFXs will be used and what they will be, and any music and sound to be used. An **AUDIO ASSISTANT** helps the AD accomplish these plans. A **SET DESIGNER** designs any set that must be built. Set construction is usually done by an entirely different person who works directly with the set designer and the director.

The **TECHNICAL DIRECTOR** makes certain that the picture being recorded is technically correct and pushes the buttons on the switcher (control box

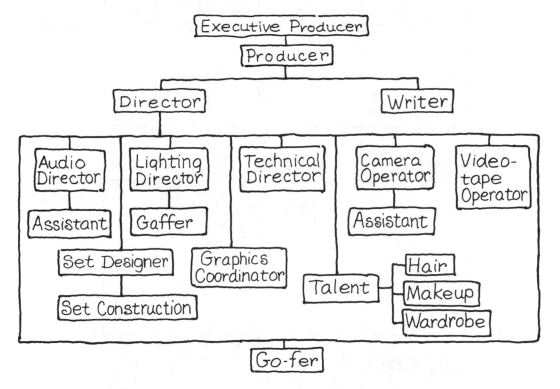

Figure 1–12 The production hierarchy: who is in charge and who works for who.

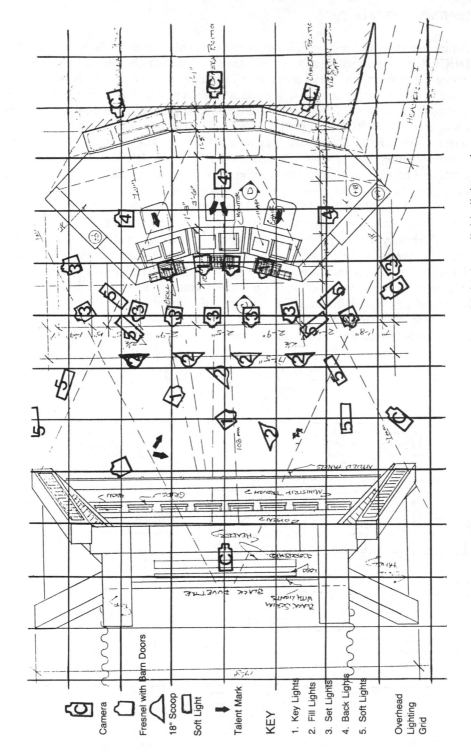

Figure 1–13 A lighting plot: a sketch of the location with both available light and any light to be added indicated in its exact location and intensity. Courtesy of JM PRODUCTION CO. FINDERS KEEPERS.

Camera

Fresnel with Barn Doors

18" Scoop

Soft Light

Talent Mark

KEY

1. Key Lights
2. Fill Lights
3. Set Lights
4. Back Lights
5. Soft Lights

Overhead Lighting Grid

for all elements in the taping) to the director's call for camera shots, effects, or pre-recorded tape. The **VIDEOTAPE OPERATOR** rolls tape as called for by the director and makes certain that the recording machines are operating properly. A **GRAPHIC COORDINATOR** makes sure all art, slides, or pictures to be used are in the right place at the right time and may also create computer graphics on the spot. A **WARDROBE** person gathers the clothing the talent will wear and the **MAKEUP** and **HAIR DESIGNERS** make sure the talent looks like the director wants them to look at the right time.

A **GRIP** does odd jobs around the shoot including laying cable or assisting the gaffer, the assistant audio director, or the camera operator. A **GO-FER** is named so because this person usually "goes for" anything that is needed on the shoot.

AVAILABLE LIGHT is what is naturally at the location, whether it is fluorescent lighting in a gymnasium or sunlight through a window. A **LIGHTING PLOT,** as shown in Figure 1–13, is the plan or design for the lighting, showing all available light and all light you are going to add. If you are asked to **TWEEK** a light, you are being told to finely adjust it to be brighter or dimmer. Lighting **HOT SPOTS** are simply areas that have too much light.

AMBIENT SOUND refers to the natural sounds at a location, such as the sounds of a freeway in the background or the sound of typewriters in an office. **MIKES** or microphones are used to transmit the sound to the videotape. **HAND-HELD MIKES** are microphones held by the talent. A mike may also be placed on a **MIKE STAND.** A **BOOM MIKE** is one held by a grip above or below the talent but out of the picture frame so it cannot be seen in the shot. A **LAVALIERE MI-CROPHONE** is small, designed to be worn by the performer without being conspicuous. For example, on the evening news you will see Lavaliere mikes pinned to the lapels of news anchors in the studio while the reporters in the field use handheld mikes.

When you are recording, you are **SHOOTING**. You may keep track of your shots on a **SLATE,** like the one shown in Figure 1–14. A slate is a board on which

Figure 1–14 The slate may be used to show the beginning of a new shot and includes all pertinent information about that shot.

you write the shot or scene number, the take number, the date, an indication of whether sound is being recorded, and sometimes the director and camera operator's name. If you use a slate, you record it on the tape immediately prior to recording the shot itself.

When you are **LOGGING A SHOT** in the manner shown in Figure 1–15, you are noting a reference point for locating that shot again, such as the counter number on the videotape recorder, cassette number, time of day, shot number, and any comments.

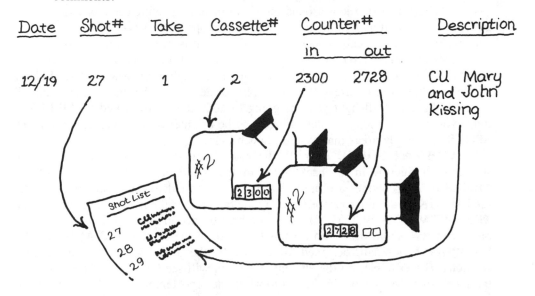

Date	Shot#	Take	Cassette#	Counter# in	out	Description
12/19	27	1	2	2300	2728	CU Mary and John Kissing

Figure 1–15 Logging a shot: making a note of all information needed to easily find that shot again.

Broadcast television uses **TIME CODE** numbers as a reference point. Time code is a series of numbers actually recorded on the tape and read by the machine as the tape plays back. Time code is a series of eight numbers which indicate the hours, the minutes, the seconds, and the frames. These are written as shown in Figure 1–16, 00:00:00:00. If a shot was located at 1 hour, 23 minutes, 13 seconds from the beginning of the tape, the time code number would read 01:23:13:00. The frame number, which is the last set of zeroes, allows you to find an exact picture. It takes 30 pictures or frames to make a second of television. In this way, video is just like film, recording a series of still pictures to make moving pictures. The difference, of course, is that you can see the pictures recorded on a piece of film while you cannot see those recorded on videotape. Time code makes it possible, however, to find one picture electronically, giving you **absolute** control of the recorded material and making it possible to edit down to a single frame.

The average camcorder does not have this fine control because it does not record a counter number on the tape. Counter numbers are purely a mechanical

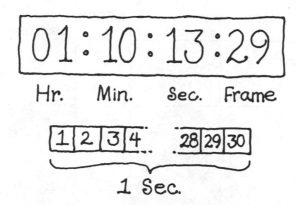

Figure 1–16   Tape time when counted in hours, minutes, seconds, and frames. There are thirty frames to one second of videotape.

function of the machine itself. They do not relate to a specific picture at all. Thus you don't have as much control as you would with time code but you can at least get close to a specific frame by keeping good counter number notes.

A **TAKE** is when you record a shot. Take 1 is the first time you record. If you record the shot again, then you log it Take 2 and so on. If you have more than one take, you will select one for the show when you edit. A **PICKUP** is when you go back to record a shot that you passed over and did not record at the time you had planned to for one reason or another. *Or* you could do a pickup on a shot that did not record well, whether the problem was quality of picture or action.

If someone tells you to **PUT A CLOCK ON IT,** they are telling you to time that particular scene. This is usually done with a **STOP WATCH,** since in television every second counts. This is particularly true in broadcast television. Correct timing is the number-one rule in broadcast television, where if the show is a second long, it gets chopped off to make room for the commercial. The commercial is, after all, where the money is made on the broadcast side. When you time a shot on your stop watch, note the minutes and the seconds as follows: 01 : 00. In broadcast television, you may see time recorded with eight numbers as is done when noting time code, 00 : 01 : 00 : 00.

When you are planning the composition of a shot, that is, what is going to be seen in it, you are **FRAMING** the shot. When the director asks for the camera operator to do a certain shot, the director has **CALLED** for that shot. The way a shot is framed, as seen in Figure 1–17, dictates how the director will call for that shot. A **WIDE SHOT** (abbreviated **WS**), for example is the widest possible shot of the location; a **CLOSEUP** (abbreviated **CU**) is a closeup of the principal action; an **EXTREME CLOSEUP** (abbreviated **ECU**) is tighter than a closeup; and a **MEDIUM SHOT** (abbreviated **MS**) is somewhere between a wide shot and a closeup. When the director calls for a **REVERSE ANGLE,** he is asking the camera operator to move to the opposite side to shoot from another perspective. You may also call for a

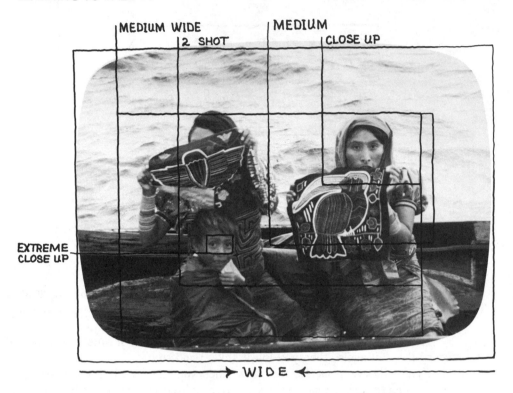

Figure 1–17 Calling for a shot by the way it is framed.

ONE SHOT, a **TWO SHOT,** a **THREE SHOT, GROUP** or **CROWD SHOT;** the number refers only to the number of people or, sometimes, principal objects in the shot.

In order to ensure that the scene has been captured completely, the director will generally plan to shoot (1) wide shots to establish location; (2) closeups to direct the audience's attention to the main action, object, or person in the scene; (3) reverse angles to show the viewer more than one side of an action, object, or person; (4) **CUTAWAYS,** which are specific closeups or extreme closeups of action, objects, or people that will be used to further the story or explain something, as shown in Figure 1–18; and (5) cover shots of anything considered visually exciting that may be valuable later in the edit to cover a taping error or problem.

A **SPLIT SCREEN** is a video EFX and means to simply split the screen into parts to insert more than one picture, as in Figure 1–19. For example, a **2-WAY SPLIT** cuts the screen in half and puts one picture on one side and another picture on the other side. A **QUAD SPLIT** cuts it into four pieces; a 16-way split into 16 different pictures. The larger the split, the smaller the pictures. A **32-WAY SPLIT,** which has been done, makes for very small pictures but the idea is not to see the

Figure 1–18 Inserting a cutaway between two shots to explain or further the action or to act as a transition.

2 Way Split   Quad Split   16 Way Split

Figure 1–19   Split screen: multiple divisions of the picture.

Live : direct to your t.v.

Live on tape: camera to tape to t.v.

Figure 1–20   An example of **live** and **live on tape** type of recording.

pictures as much as it is to create the effect. Splits are generally done with large switchers with effects capability. You can, however, simulate this effect with mirrors or a bank of television sets with a little imagination.

When a show is **LIVE,** it is being recorded as it happens, with or without the benefit of a script, as demonstrated in Figure 1–20. The *Muscular Dystrophy Telethon* is live. A show recorded **LIVE ON TAPE** is simply a show recorded as it happened but aired at a later time, such as *The Tonight Show.* A show recorded live on tape may have been edited. A show that is **AIRED LIVE** is aired exactly as it happens with no editing. **PORTIONS PRERECORDED** means that parts of the show have been recorded at another time and may have been edited. This prerecorded footage is inserted at the appropriate time in the live show. For example, during your local evening news, you often see reporters in the field do live introductions for prerecorded stories.

If you are taping a sequence exactly as it happens with no editing at all, you are recording in **REAL TIME.** For example, if you were taping a wedding and rolled taped for the entire ceremony and reception leaving nothing out, you would have a real-time-video. You would also have a video about three hours long, which would be ponderous to view, and thus the reason for the edit, to cut the show down to only that pertinent footage that will tell the story.

A camera can be **HANDHELD** or **MOUNTED.** A handheld camera is carried by the camera operator. A mounted camera is attached to something, usually a tripod sometimes called **STICKS,** to give the camera more stability. If the tripod is on a flat surface that has wheels, the camera is said to be on a **DOLLY,** which allows it to move without giving up the stability of the tripod. See Figure 1–21.

Figure 1–21   The ways cameras are used on a shoot.

A camera can be **LOCKED OFF,** meaning it is locked into a particular shot and has no operator and thus is not capable of any movement, whether lens movement or physical movement. A camera might be locked off on a wide shot while a second camera records closeups of the action with the idea that with the wide shot on tape, you always have the option of cutting there if you need to. It just gives you more flexibility when you edit. A camera can be **FLOATING,** meaning it has no particular position, but rather moves at the operator's will or at the director's call, from position to position.

A camera operator may be asked to **TILT UP** or **TILT DOWN,** meaning to actually tilt the camera as directed; to **PAN LEFT** or **PAN RIGHT,** to turn the camera to the left or right with the camera operator staying in the same position. To **TRUCK LEFT** or **TRUCK RIGHT** means to physically move the camera to the left or right by pushing the dolly or moving the tripod on which it is mounted to the new position. Camera movements are demonstrated in Figure 1–22.

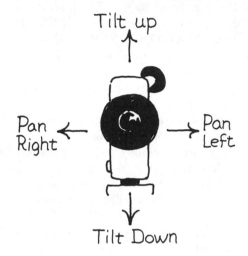

Figure 1–22   How a camera can physically move.

To **PULL** or **PUSH,** also sometimes called for by saying **ZOOM OUT** or **ZOOM IN,** as seen in Figure 1–23, means doing just that using the camera lens. If you are asked to **LOOSEN UP** on the shot, you are being told to zoom out just a bit more. If you are asked to **TIGHTEN** the shot, you should zoom in just a bit.

FOCUS means just that, to focus the camera. **SPLIT FOCUS** means to focus the camera *between* more than one subject so that neither will be radically out of focus, but rather the main focal point will be between the subjects, giving them all the appearance of being in focus. **RACK FOCUS** is to go slowly out of or into focus. A rack focus is often used as a simple video EFX or as a transition to the next scene. If

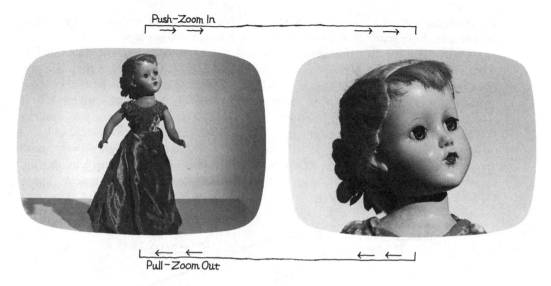

Figure 1–23   Camera movement using the lens.

you are **SOFT,** you are out of focus. You may deliberately shoot soft for effect or to make someone look better. You can also achieve this soft effect with lighting or filters if preferred.

A **LOCATION** is anywhere you plan to tape all or part of your show. You may have multiple locations. As shown in Figure 1–24, locations fall into two basic categories: (1) **EXTERIORS,** outside shots, or (2) **INTERIORS,** inside shots. Your location may be in a **STUDIO,** a warehouselike room designed specifically for TV or film production. If you are taping away from the studio, you are **ON LOCATION** taping **IN THE FIELD,** sometimes referred to as **REMOTES.** The late-night news,

Figure 1–24   Locations are either studio or field, interiors or exteriors.

for example, has a studio location for the anchors and interior and exterior remotes. Sometimes these remotes are live; sometimes prerecorded on tape.

When you are directing your production, you need to know basic **STAGE DI-RECTIONS** shown in Figure 1–25. Looking toward the scene from the camera's perspective, the area closest to you is **DOWNSTAGE.** The area furthest away from you is **UPSTAGE.** When the actor is facing the camera, the area to his left is **STAGE LEFT.** The area to his right is **STAGE RIGHT.** When you are looking through the camera or at the scene on a TV, you will see this in reverse. Right will be left and left will be right. It is, therefore, important to remember these basic stage directions when you are **BLOCKING** the show, which is simply telling the talent what to do, where to go, and when. All the area not in the camera shot is **OFFSTAGE** or **BACKSTAGE.**

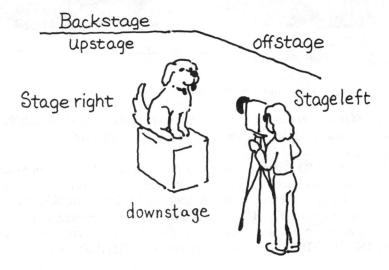

Figure 1–25   Stage directions. These are used to tell the talent where and in what direction you want them to move.

The **EDIT** is when you put all the shots together to make the show. An edit may be complicated with many **SOURCES,** which are what you call the playback machines in the edit, or it can be a basic **TWO-MACHINE EDIT,** meaning you have one machine on which to record and one machine on which to play back. Some edit configurations are shown in Figure 1–26. Most camcorders can be used as play-back machines, and thus can become one source. Some basic edit systems allow you to do **CUTS ONLY EDITING,** meaning that you can do nothing but cut from one scene to the next. You have no dissolve, wipe, or other video effects ability. If the system is not **FRAME-ACCURATE,** you cannot edit to the exact frame. You will remember that television is recorded at 30 frames per second. A system that is not frame-accurate will mean that your editing will not be perfect to a single frame, but rather the video and audio will slip forward and backward with each edit. This type

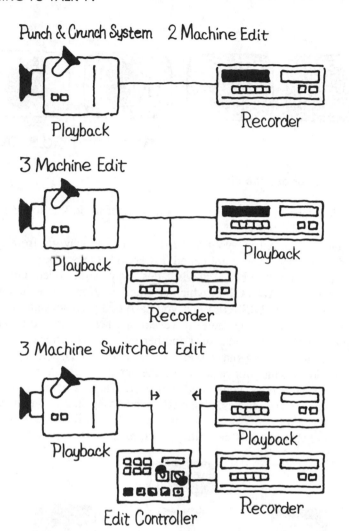

Figure 1–26 Some types of home edit systems.

of system is sometimes called a **PUNCH AND CRUNCH EDIT SYSTEM.** You can add a small **SWITCHER** which is just an edit controller that will allow you more accuracy *and* allow you to do fancier edits, or you can use a more technically sophisticated VCR that has some edit capabilities built into it.

A **SINGLE CAMERA SHOOT** is taping a show with one camera only. A **MULTIPLE CAMERA SHOOT** is taping with more than one camera. If you add a switcher to a multiple camera shoot, you will have the ability to switch from one camera to another as you are recording, thus minimizing or eliminating the need for an edit. A **ONE CAMERA-NO EDIT SHOOT** is simply taping with one camera with no plans to edit, but rather you are totally committed to what is recorded on the

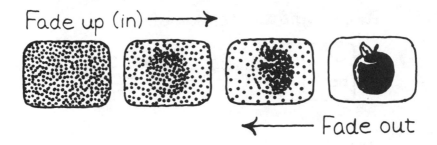

Figure 1–27   Fading in or out of a picture.

piece of videotape in the camera. You might elect to do this to save time or if you have no edit capabilities at all.

When you are putting your show together, you may get from one shot to another in several ways. These are called **TRANSITIONS.** You can **FADE UP** or **IN** from black, meaning the black slowly fades away as the video and/or audio slowly comes in. You can **FADE OUT,** which is simply a reversal of a fade up, as shown in Figure 1–27. You can **DISSOLVE** if you have two playbacks and a switcher, which means that one picture fades away to reveal another picture. You can **CUT,** which means you end one picture and start another immediately after it. If you **WIPE,** you are bringing in or taking out a scene with an EFX that actually wipes the scene away. It is like wiping your hand across the screen, with the picture going away or coming up as you move your hand across it. There are many different kinds of wipes including stars, circles, diagonals, and even fancy combinations. Some of these are shown in Figure 1–28. Wipes can be done at the time of taping if you have the equipment that has this capability, such as a VCR, a switcher, or a **SPECIAL**

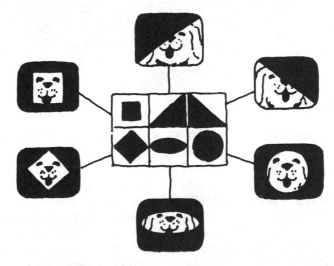

Figure 1–28   Wipes: bringing in or taking out pictures with an effect.

EFFECTS GENERATOR (SEG), a piece of equipment that allows you to cut between cameras or add visual effects. Wipes can also be done in the edit using an edit controller or an SEG that has wipe effects built into it.

A **CHROMA KEY,** sometimes called just a **KEY,** is when you put two pictures together to make one. If you want to ski, for example, but you do not want to bother going to the slopes, you could key a shot of yourself over a shot of the mountain and you would be there. Keys are common in the weather portion of the broadcast news. The weatherperson stands in front of a blue wall or board and the weather picture is keyed into the blue as shown in Figure 1–29.

Figure 1–29   Chroma key: putting one picture over another picture. Photo courtesy of Jim Barach and KCOY-TV Santa Maria, CA.

A **SUPER** is like a key but usually refers to graphics and titles over a picture instead of a picture over a picture. Titles and credits are often supered over opening or closing shots as shown in Figure 1–30.

The **CREDITS** are a list of the names of the people who worked on the show and may include the director, the writer, the performers, and the camera operators, among others. Credits are sometimes called **CHYRON®**, which refers to the fancy

TITLE ONLY   SUBTITLE

Figure 1–30   Titles and subtitles on the show may be supered over black or action.

computer made specifically for the purpose of putting credits or other graphics on broadcast TV shows. Chyron® has become synonymous with credits, much as Xerox® has become synonymous with photocopying. **CHARACTER GENERATORS** have the basic ability to generate credit lists provided they are words only. Credits may be followed by **THANK YOUS,** a list of people or places that helped in some way but were not directly involved in the production.

The **OPEN** is all that appears at the beginning of the show. It may be just the **TITLE** of the show or a montage of shots of the performers or the action or a specially edited tease. A **TEASE** is a sequence at the beginning of the show that is an abbreviated version of the show itself or a piece of the show. It is designed specifically to capture the attention of the viewer and entice them to keep watching. You can see an example of a tease by turning on your television set any night and catching the first video of your favorite show. Generally what you will see is various scenes from that night's show. They are trying to keep you from changing the channel by capturing your attention with a show tease. A title may also include a **SUBTITLE** such as a location or time of day, as shown in Figure 1–30.

The **CLOSE** is the ending of the show, usually the video and audio following the last spoken word. For example, the cowboy gets on his horse and rides off into the sunset. The credits are a part of the close.

BUMPERS are isolated pieces of the show that are inserted as one piece prior to a commercial break or at the end of the show. A bumper may be a tease for something to come, a trivia question, a visual recap of what the audience has been watching, or just the name of the show supered over a visual with a voice over saying, "We'll be right back." On the late-night news you will see bumpers right before the commercial saying, "coming up next . . ." and then a voice over description with a visual of a story that will be featured in the next segment of the news after the commercial.

Figure 1–31 Technical difficulties: all those problems that can and probably will happen on a television shoot.

THAT'S A WRAP means you have ended the taping day and when you **STRIKE,** you are packing up your equipment or tearing down a set or removing props or removing anything else you brought to the taping site.

And finally, **TECHNICAL DIFFICULTIES,** originally meant to refer only to equipment problems on a live show, is now a catchall phrase for virtually *anything* that can and *will* go wrong on the shoot. We have included here in Figure 1–31 a sign just for you to insert over your technical difficulties.

CONCLUSION

This is only some of the TV Talk you will hear in the business. We are sure you will even coin a few of your own words as you produce your videos. Television is never the same. It is totally unpredictable and there are always new situations demanding the creation of new words to describe them. As long as there are new ideas for shows and new people producing them, there will be new words.

With the TV Talk you have just learned, you now know what it means to choose your videotape; write a script, format or storyboard; watch out for bleeding; keep your tape generations to four; roll tape; burn a take; create a video or audio EFX; cast your principal talent; hire your go-fer; tweek a light; place a boom mike; log a shot; put a clock on a shot; call for a WS; do a quad split; put a camera on sticks; float a camera; rack focus on a shot; block a principal to stage right; chroma key an effect; roll credits; tease the open; add bumpers; and wrap the shoot.

Coming up next, *The Pieces to the Plan,* or how to plan your way to a great video.

CHAPTER TWO

THE PIECES TO THE PLAN

Making television is fun—at least that is what everyone says. Actually, *it is* fun but it is also a lot of paper and planning. It may seem sometimes like the planning–taping ratio is about 100 to 1, but don't let that fact scare you off, because even the planning can be fun—at least we are going to *make* it fun.

WHAT IS THE KEY TO A GOOD VIDEO MOVIE?

Knowing what you are shooting, where you are going, **and** how you are going to get there **before** you roll tape is the true key. And all that means is devising a good plan and executing it.

A Good Idea

First, the idea. Is it a good one? Will it make a good video? Will the audience you are making it for want to watch it? Does anyone care? Do you care? Will you watch? If you can truthfully answer yes to these questions, then you may have a good idea and you may not. That is the way television is. You never really know until you know and even then you may not know for sure. Make sense? No? Well, it is not intended to. It was only intended to get you to ask yourself if your idea is a good one.

Birthdays are good. Weddings are good. A new baby is good. Even your own version of a TV game show or your favorite movie scene are good. Or if you are making it for your business, a new product or a convention or a marketing strategy are good ideas. But are any of these good enough in themselves to make a good video? That is, a video that anyone will watch more than once?

Possibly, if you plan it right. For example, if you include in that video something other than just shots of, say, the wedding, like Uncle John napping on the kitchen floor after hitting the punch bowl several times too many or an interview with the "old" boyfriend of the Bride's or photos or other video of the happy couple on their first date or a rundown of the family tree on both sides including the great great uncle hanged for horse stealing. This type of thinking beyond the obvious will make the video unique and thus memorable, just as interesting and exciting in rerun as it was the first time around. Then and only then, you may have yourself a **good** video.

Don't misunderstand. You do not need to bend the idea to make it interesting and exciting. You don't need to dream up things. You just need to think it through beyond the obvious content and make sure that you can create a show that both achieves your goal *and* does it in a way that will make the audience want to watch. A wedding doesn't have to be cute to be a good video. You could add the right music, the right visuals and make it a truly touching experience causing everyone that watches it cry with happiness. That is success: to touch the audience and make them feel or care about something whether their reaction is laughter or sadness or just genuine interest.

The idea, then, is important. No—it is *critical* to the success of your video. If you do not feel that your idea is strong enough, then either abandon it entirely or make it better. If, for example, the wedding is expected to be conventional, then add a bit of excitement. Perhaps a car chase as the happy couple drives off into the sunset. This can be easily done through the *magic of television* without actually intruding on the happy moment. Use your imagination. Remember the *A Team*. That was a simple idea made great with gimmicks and chases. You may not have thought it was an intellectual success but you watched. It was one of the highest rated shows on the air at one time. That is, after all, all you want to do—to please your audience, to get them to come back for more. It all begins with the idea.

Planning

Television is planning, planning, and still more planning. How good your video finally is will be largely dependent on how good your plan was and how well you execute it. In creating a good plan, you are going to want to:

1. Choose your subject
2. Create the show
3. Select your performers
4. Plan the production details
5. Tape the show
6. Review the taped material
7. Put the show together
8. Have your **PREMIERE** showing

In the chapters to come we will share some planning techniques with you that will help you create the best possible video. We will cover each of the following in depth in this book.

Goal Set.   One of the first things you want to do is to set your goal. Define exactly what you want to do, whether it is making a romantic wedding video or creating excitement for a new product. Whatever your goal, *set it* and do your best to *achieve it*. For it is that goal that will create the pictures and sound that will make a successful video.

Good Lists.   Television runs on paper. As a matter of fact, sometimes it may seem like there is more paper than videotape. Making lists is, therefore, a big part of the planning process. With comprehensive, workable lists, you can keep your troubles to a minimum, your organization to the max, and your show a number-one production. The lists we will cover in this book include:

1.   MASTER CHECKLIST (see Figure 2–1).   For use throughout the production. A comprehensive list of all elements of the taping. You will use this to make certain that you have completed all aspects of preparing for the taping including the gathering of all the material you need for that taping.
2.   OUTLINE (See Figure 2–2).   For preproduction use. A basic overview of what you plan to tape, what equipment you will use, and when you will tape. It is the first step in turning an idea into a viable subject for a video.
3.   SCHEDULE (See Figure 2–3).   For use throughout the production. The complete time schedule for the shoot, including the preproduction, the production, and the postproduction. This schedule will map out the day-to-day activities during the entire production cycle from the initial idea to casting, to the formatting and scripting, to the location-scouting to the taping sessions, to the final edit.
4.   BUDGET WORKSHEET (See Figure 2–4).   For preproduction use. This worksheet is designed to let you make a best-guess estimate of what the video will cost including all possible expenses and fees.
5.   EQUIPMENT LIST (See Figure 2–5).   For preproduction and production use. A complete list of all the equipment you will need for a professional caliber shoot working within the limitations of what is realistically available.
6.   CREW LIST (See Figure 2–6).   For preproduction and production use. A complete list of the crew and their duties.
7.   CAST LIST (See Figure 2–7).   For preproduction and production use. A complete list of all cast members by role and real name.
8.   LOCATION LIST (See Figure 2–8).   For preproduction and production use. A complete list of all locations and the dates and times of the scheduled taping at those locations. This may also include the shot list for the location, if you wish.
9.   PROP LIST (See Figure 2–9).   For preproduction and production use. A list of all props you will need on the shoot, noting the particular location where they will be needed.

Show Title __Western Shootout__

Date __12-5__

MASTER CHECKLIST

(Add or delete items as appropriate to your show)

(Check when completed)		(Date completed)
✓	OUTLINE	12-5
✓	BUDGET	12-5
✓	EQUIPMENT LIST	12-5
✓	CREW LIST	12-5
✓	SCENARIO	12-6
✓	FORMAT	12-6
✓	SCRIPT	12-6
✓	STORYBOARD	12-6
✓	CAST LIST	12-6
✓	MAKEUP	12-7
✓	WARDROBE	12-7
✓	PROP LIST	12-7
✓	LOCATION LIST	12-7
✓	SET DESIGN	—
✓	SHOT LIST	12-8
✓	SCHEDULE	12-8
✓	LOCATION SKETCHES	12-7
✓	CAMERA POSITIONS	12-7
✓	LIGHTING PLOT	12-7
✓	BIBLE	12-9
✓	SHOT LOG	12-10+
✓	CREDITS	12-17
✓	EDIT LIST	12-17
✓	DUBS	12-20

Figure 2–1 Master Checklist.

SHOW OUTLINE

DATE: __12-5__

SUBJECT: __Western Shootout__

PURPOSE: To document _____   To entertain __X__

To inform _____   To motivate _____

To make the audience happy __✓__   Sad _____   Think _____

ROLES: __Marshal Goodguy__   TALENT _____

__John Sleeze__   TALENT _____

__Mary Sunshine__   TALENT _____

__Jerry Lee__   TALENT _____

__Bartender__   TALENT _____

__Townsfolk__   TALENT _____

THE STORY

THE PLOT: __John Sleeze, the bad guy, comes to town to take on__
__the Marshal.__

SUBPLOT: __Mary tries to persuade the Marshal to leave.__

THE CLIMAX: __Shootout between Sleeze and the Marshal.__

SUBCLIMAX: __Marshal quits__

THE CLOSE: __Mary and Marshal leave town.__

CREDITS: yes __✓__   no _____

Over close __✓__   black _____   show scenes _____

THE OPEN: over black _____   story __✓__   action sequence _____

What does it look like? __sleepy Western town__

THE STYLE

THE PICTURES: __dramatic - intense - closeups__
__action__

Figure 2–2 Show Outline sample.

THE MUSIC: yes __✓__   no _____   name it _Western - dramatic_

instrumental __✓__   lyrics _____

THE SOUND EFFECTS: yes __✓__   no _____   what? _clock ticking_

SPECIAL ELEMENTS: slides _____   snapshots _____   film _____

video _____   graphics/artwork _____   special effects __✓__

what? _what ? slow motion_

THE LOCATIONS: _Red Dog Saloon_ ____ DATE _2-10_

Streets of San Ramon ____ DATE _2-11_

Marshal's Office ____ DATE _2-11_

_____ DATE _____

_____ DATE _____

_____ DATE _____

THE LENGTH: less than 3 min. _____   3 to 5 min _____

5 to 10 min _____   10 to 15 min _____   15 to 20 min __✓__

over 20 min (specify) _____

SCREENING DATE: _March 15_

Figure 2–2 (*cont.*)

SCHEDULE

DATE SCHEDULED	EVENT
12-5	BUDGET COMPLETE
12-5	EQUIPMENT CHOSEN
12-5	CREW CHOSEN
12-6	FINAL SCRIPT APPROVAL
12-6	CASTING
12-7	LOCATIONS SELECTED
12-7	WARDROBE, MAKEUP AND PROPS SELECTED
12-8	PREPARE SHOT LIST
12-8	SCHEDULE TAPING DAYS
12-7	POSITION CAMERAS AND LIGHTING
12-10 thru 12-16	TAPING
12-17	MAKE EDIT DECISIONS
12-18	EDIT
12-20	DISTRIBUTE DUBS

Figure 2–3   Schedule sample.

Show Title _Western Shootout_

Date _12 − 5_

BUDGET WORKSHEET

Note: Estimate all costs. It is not necessary to make any final decisions about locations, equipment, crew, edit, or set at this time. This budget is a best-guess estimate of all expenses for the proposed video.

VIDEOTAPE

Cost per tape	$ 5.98
No. Cassettes	× 9
Total	$ 53.82

LOCATION FEE

(Place, if you know)

_____	$ _____
_____	_____
_____	_____
Total	$ ⊖

TRANSPORTATION

Gas	$ 30.00
Parking	15.00
Toll charges	5.00
Airline tickets	⊖
Misc.	20.00
Total	$ 70.00

PROPS

Buy	$ ⊖
Rent	$ 15.00
Total	$ 15.00

Figure 2–4   Budget Worksheet sample.

EQUIPMENT

(Name it, if possible)

Video _____

_____

_____ $ _____

Audio _____

microphone

20.00

Lighting _____

Pro light

10.00

Misc. _____

_____

Total $ 30.00

CREW

Cost per $ 10.00

No. crew members   × 6

Subtotal   60.00

Plus other   + _____

Total $ 60.00

EDIT

Rent/buy equip _____

_____ $ _____

Subtotal $ —0—

Postproduction facility

cost per hour $ 10.00

No. hours est.   × 2

Subtotal $ 20.00

Total $ 20.00

Figure 2–4 (*cont.*)

TELEPHONE

    Estimated cost $ 5.00 $ 5.00

ARTWORK/GRAPHICS

    Misc. expenses $ 10.00

    Artist   0

               Total $ 10.00

CATERING

    Estimated cost $ 20.00 $ 20.00

MUSIC

    Buy records/tapes $ 8.00

    Original music $

               Total $ 8.00

ADDITIONAL VISUALS

    Photos $

    Video   3.00

    Film

    Other

               Total $ 3.00

POSTAGE/DELIVERY SERVICE

    Stamps $ 10.00

    Delivery $

               Total $ 10.00

PHOTOCOPYING

    Estimated cost $ 30.00

    Supplies

               Total $ 30.00

SET

    Planning $ 0

    Construction

               Total $ 0

Figure 2–4 *(cont.)*

CAST

    Expenses $ 20.00

    Per diem

                Total $ 20.00

DUB COST

    Cost per tape $ 5.00

    No. cassettes   ×   10

    Subtotal $ 50.00

    Dubbing fee $ 5.00

    No. cassettes   ×   10

    Subtotal $ 50.00

                Total $100.00

WARDROBE

    Buy $ 0

    Rent   30.00

                Total $ 30.00

MAKEUP/HAIR

    Supplies $ 20.00

    Misc. expense $ 0

                Total $ 20.00

TOTAL ALL $504.82

MISC. EXPENSES

    Total all $ 504.82

    Add 10%   ×   .10

                Total $ 50.48

YOUR FEE $ 0

TOTAL BUDGET $ 555.30

Figure 2–4 *(cont.)*

Show Title __Western Shootout__

Date __12-5__

EQUIPMENT LIST

CAMERAS

Total no. __3__

Format (specify number)

VHS __2__   Beta _____   8 MM __1__

Designated duties

#1 __Action - angle #1__

#2 __Action - angle #2__

#3 __Reaction close ups__

#4 _____

MICROPHONES

Total no. __1__

Type (Specify number of each)

Handheld _____   Boom __1__

Mike stand _____   Lavaliere _____

LIGHTING

Total no. units __3__

Type (Specify number)

Pro lamps __1__   Shiny board __1__

Camera mounted __1__   Other _____

Figure 2–5   Equipment List sample.

TAPE

Total no. cassettes _____ 9 _____

Kind (Specify number)

VHS ___ 6 ___   Beta _____   8 MM ___ 3 ___

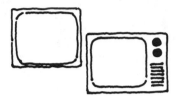

TVs/MONITORS

(Number and size of screen)

No. Black & White _____

No. Color _____ 2-19" and 5" _____

ACCESSORIES

(Specify number)

Tripods/stands

Camera ___ 1 ___   Lighting ___ 1 ___

Audio _____   Dollys ___ 1 ___

Lighting gels (color and number of each)

color ___ 1 ___   # *orange*

color ___ 1 ___   # blue

VCRs

Total no. ___ 1 ___

Kind (specify number)

VHS ___ 1 ___   Beta _____   8 MM _____

Figure 2–5 (*cont.*)

Show Title **Western Shootout**

Date **12 – 5**

CREW LIST

DIRECTOR __you__

LIGHTING DIRECTOR __you__

TECHNICAL DIRECTOR _____

AUDIO DIRECTOR __you__

SET DESIGNER _____

SET CONSTRUCTION _____

VIDEOTAPE OPERATOR _____

CAMERA OPERATOR #1 __Tom__

CAMERA OPERATOR #2 __Sandy__

CAMERA OPERATOR #3 __Robert__

CAMERA OPERATOR #4 _____

GAFFER __Edward__

GRIP __Roland__

MAKEUP ARTIST __Michelle__

HAIR STYLIST __Michelle__

WARDROBE __Michelle__

PRODUCTION ASSISTANT __Marie__

_____

ASSISTANT TO:

CAMERA 1 __Ralph__

CAMERA 2 __James__

CAMERA 3 __Melissa__

CAMERA 4 _____

GO-FER __Ronnie__

Figure 2–6 Crew List sample.

Show Title _Western Shootout_

Date _12 - 6_

CAST LIST

ROLE _Marshal Goodguy_   TALENT _Joe_

ROLE _John Sleeze_   TALENT _Jonathan_

ROLE _Mary Sunshine_   TALENT _Sue_

ROLE _Jerry Lee_   TALENT _Brian_

ROLE _Bartender_   TALENT _Mikael_

ROLE _Sleeze man #1_   TALENT _Jim_

ROLE _Sleeze man #2_   TALENT _Robert_

ROLE _Mayor_   TALENT _Earl_

ROLE _Undertaker_   TALENT _Bob_

EXTRAS _Townsfolk_

Figure 2–7   Cast List sample.

Show Title __Western Shootout__

Date __12 - 7__

LOCATION LIST

Locations:
Interior   Exterior

FIELD LOCATIONS	EXTERIORS	INTERIORS
Red Dog Saloon	✓	✓
Streets of San Ramon	✓	
Marshal's Office	✓	✓

STUDIO LOCATION		
Director's House		✓

Figure 2–8 Location List sample.

Show Title _Western Shootout_

Date _12 – 7_

PROP LIST

ITEM	NEEDED AT LOCATION
1 Rifle	Marshal's Office
1 Rifle	Red Dog Saloon
4 Western Gunbelts w/guns	All
Clock	Red Dog Saloon
Clock	Marshal's
Badge	All

Figure 2–9   Prop List sample.

Show Title ___Western Shootout___

Date ___12-8___

SHOT LIST

LOCATION ___Red Dog Saloon___

Note: I = Interior   E = Exterior

Framing: CU, MS, WS, ECU, MCU, etc.

I or E	SHOT #	FRAMING	DESCRIPTION
E	22	WS	Saloon
I	23-1	WS	Inside Saloon
I	23-2	CU	John Sleeze
I	23-3	CU	Bartender
I	23-4	M 25	Two men at table
I	23-5	MW	Sleeze at bar

CALL TIME ___9 AM___

DAY ___December 10___

CALLED CREW (if all, just say "ALL")

___All___

CALLED CAST (if all, just say "ALL")

___John Sleeze___

___Two Townfolk___

___Bartender___

Figure 2–10 Shot List sample.

Show Title __Western Shootout__

Date __12 - 10__

Producer __You__

Location __Red Dog Saloon__

SHOT LOG SHEET

Page __1__ of __2__

Camera No. __1__

Camera Operator __Tom__

Camera Assistant __Ralph__

COUNTER NUMBER

CASSETTE #	SHOT #	TAKE #	IN	OUT	APPROX TIME	NOTES
1	22	1	0020	0070	:30	Static includes townfolk
		2	0080	0130	:30	Static without people –
1	23-1	1	0140	0345	2:05	choppy –
		2	0355	0505	1:50	good – needs cover at gunshot
		3	0515	0672	1:57	great – use this one
1	23-5	1	0682	0692	:20	Sleeze needs to be tougher
		2	0702	0729	:27	good – cover with CU
1	23-2	1	0739	0749	:10	tough cover for 23-5 Take 2

Figure 2–11 Shot Log sample.

Show Title _Western Shootout_

Date _12 - 17_

EDIT LIST

CAMERA #	CASSETTE #	SHOT #	COUNTER NUMBERS	CONTENT NOTE
Music	Up under shot	#22		
1	1	22	0080 - 0130	WS Saloon
2	1	23 - 1	0257 - 0315	bar WS
3	1	23 - 6	0152 - 0171	CU Sleeze hand hitting bar
insert { 1	1	23 - 1	0140 - 0200	MS scene
3	1	23 - 3	0252 - 0262	CU bartender
2	1	23 - 2	0654 - 0671	CU Sleeze
continues		23 - 1	0575 - 0600	action
3	1	23 - 4	0315 - 0330	2 men at table

MATERIALS TO TAKE TO THE EDIT

Special art: _none_

_____

_____

_____

End credits: _yes_

Title cards: _yes_

Figure 2–12   Edit List sample.

10. SHOT LIST (See Figure 2–10).   For preproduction and production use. A complete list of all the shots and any extra sound you want to record on the videotape. This list will make it possible to go into postproduction knowing that you have recorded every picture and sound on the videotape that you will need for a successful edit session. You will not be faced with missing or bad shots that would force you into editing in mistakes that will distract from the intent of the video.

11. SHOT LOG (See Figure 2–11).   For production and postproduction use. Everything you record on tape will be noted on this log with its exact position on the tape, as close as you can get to exact given the equipment's capabilities. The log includes the time-code number or the counter number, the cassette number (if the material is recorded on more than one cassette), the shot number from the shot list, an approximate time length, and notes on the shot. With this information, you will be able to locate everything recorded when it comes time to edit your video.

12. EDIT LIST (See Figure 2–12).   For postproduction use. The edit list is basically your editing plan. From the shot log and working with the format, script, or storyboard, you will be able to prepare a list of everything you want to do in the edit session *before* you actually go to the edit. This will shorten your editing time and should make a better show, not to mention save you lots of money if you are paying for an edit suite. Editing time, if you must rent it, is the most expensive part of a production.

You may or may not need to prepare **all** these lists, depending on the complexity of your production. You will be the best judge of that. The preparation and use of these lists will be demonstrated in later chapters. For now, the form for each is shown here for your understanding of their purpose.

Final Written Show Form

To these lists, you *must* add one of the key ingredients—the final form of your show, whether it is a format, a script, or a storyboard. You should not begin to roll tape without:

1. FORMAT.   This is an informal plan for the show, a general narrative of what you see and what you hear, from the opening to the closing, with as many specifics of what is going to happen as possible. A format is flexible enough to change as you add or delete ideas before or during the taping. If no script is written, the format *must* be much more specific and less informal to make certain that the direction of the show is established and that all material is taped. A format is shown in Figure 2–13.

2. SCRIPT.   A formal plan of exactly what is to be said and done in a show. Generally, a script is written to make certain that everything, both pictures and sound, is recorded exactly as it was planned. Sometimes a script is informal, setting the mood and direction but allowing the performers room to interpret the dialogue, to improvise the dialogue and actions, to record live action as it happens, or to build

Show Title _Western Shootout_

Date _12 - 6_

FORMAT FORM

FORMAT NO. _1_   (Same as scenario no.)

DESCRIPTION: _John Sleeze enters bar -_

Shoots towns person

SCENE LOCATION: _Red Dog Saloon_

INTERIORS: Yes _✓_   No _____

Where? _Main Bar_

EXTERIORS: Yes _✓_   No _____

Where? _Front_

SPECIAL

MUSIC: Yes _____   No _✓_

What? _____

SOUND: Yes _✓_   No _____

What? _gunshots_

LIGHTING: Yes _____   No _✓_

What? _____

SET: Yes _____   No _✓_

What? _____

PROPS: Yes _✓_   No _____

What? _Clock - 3 Western gunbelts_

SPECIAL ELEMENTS: Yes _✓_   No _____

What? _Slow motion as shot man falls._

Figure 2–13   Format sample.

SHOTS TO TAPE:

Scene No.

22	1.	WS Saloon front
23-1	2.	WS Inside Saloon – entire scene
23-2	3.	CU John Sleeze – cutaway material
23-3	4.	CU bartender – delivers all lines - cutaway material
23-4	5.	M2S –2 men at table – deliver all lines - cutaway material
23-5	6.	MW – Sleeze at bar – shooting sequence – cutaway material

TAPING DATE _December 10_

Figure 2–13 (*cont.*)

a scene or the show around interview sessions. There are two different ways to write a formal script. A sample of both ways is shown briefly in Figure 2–14.

You may or you may not need a script. Sometimes a format is sufficient. Generally, the rule is: If you want to formalize all or part of the dialogue, that is, if you want to make sure that certain words are said, then you need some form of script.

3. STORYBOARD. A pictorial version of the show. A sample is shown in Figure 2–15. These are usually hand-drawn pictures of either every shot or of the major scenes in the show. Storyboards are used to demonstrate certain camera angles and style and are used to make sure that everyone from writer to director to camera operator is seeing the same thing before first tape is rolled on the video.

Execution

What then is the key to a good video? A good idea, good planning, and last but not least, execution, having the knowhow to pull it all together in the best possible way.

The execution of the plan is a true skill that is acquired through practice and experience. The more videos you make, the better you will become at creating good ones. Part of that experience is born in learning to **Think TV.**

STYLE #1

POPCORN FALLS BACK DOWN IN
EXTREME SLOW MOTION.

MUSIC: PAUSE A BEAT

POPCORN POPS IN RAPID SUCCESSION,
IN REAL TIME. NO SOUND HEARD.

> ANNOUNCER
>
> (Delivered in the style of the voice over
> in the beginning of *The Lone Ranger* TV
> show) From out of the west in a puff
> of smoke comes . . .

SOUND: SLOWLY BRING UP THE SOUND
OF THE CORN POPPING

MUSIC: SLOWLY BRING IN THE NEXT
PART OF THE MUSIC.

> a warm friend who believes in truth,
> justice, and the American way . . .

STYLE #2

SOUND: FRONT DOOR SLAM

THE MARSHAL CLOSES THE CELL DOOR BUT DOES NOT LOCK IT.

> MARSHAL
>
> (to himself) See you later, Jerry Lee.

MARSHAL WALKS BACK TO HIS OFFICE.

CUT TO OFFICE. A SWEET, INNOCENT LOOKING YOUNG WOMAN IS THERE,
WAITING PATIENTLY. SHE IS VIRGINAL IN APPEARANCE, EVERYTHING THAT
IS GOOD AND SWEET IN LIFE. THE MARSHAL ENTERS, STOPS ABRUPTLY.

Figure 2–14   Scripts: two different ways of writing one.

MARSHAL

Mary, I didn't hear you come in.

MARY SUNSHINE

You were busy. I didn't want to disturb you.

MARSHAL

(matter of fact) Right. What is it I can do for you?

Figure 2–14 (*cont.*)

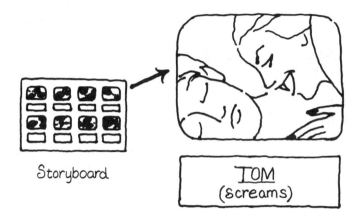

Figure 2–15   Storyboard sample.

HOW DO YOU LEARN TO THINK TV?

When you think of television, you probably think of the picture, not the sound. You might say that, "of course, the sound goes with the picture." That is true but it is also not true. If someone is speaking, certainly the sound goes with the picture. If a car drives by, the sound goes with the car. If a baby cries, the sound goes with the baby. All true. *But* what about the music? What about the wind blowing in the trees? What about the other people in the restaurant? What about sounds that are not part of the original recording? What about sound behind a still picture? What about those? Do they just happen? No, for the most part, they are planned either as a part of the original script or as the show is taped or as it is edited.

Sound is more than what you record while you are rolling tape. Sound is violins playing during the love scene or the sound effects you add because, maybe the baby crying didn't record well or you just have a better baby cry. Or maybe your performer cannot speak in the conventional sense but rather makes sounds like R2D2 in *Star Wars*. The point is that television is *both* pictures and sound and you must make your plans accordingly. You must both see and hear when you are visualizing your show.

When you think sound, think beyond the spoken word. Think of music and special sounds. Music, especially, can have a great impact on your video. What would *Star Wars* have been without the music and the sound effects? Not much. But with them, it was a classic. As important as sound was to *Star Wars,* it is just that important to your TV show.

Think of TV as two pieces, one the pictures and the other the sound. To help you to start thinking of the pieces that make up the show, try this little exercise. Get a piece of paper and write down the sound and pictures that are a part of your reality right now.

For example, ours would look like Figure 2–16 and read: **PICTURE:** OFFICE WITH TWO DESKS. THREE-TIERED BOOKCASE TO THE SIDE OF THE DESK. A LOW TABLE WITH TV, VCR AND STEREO. A WINDOW OPEN OVER THE DESK. ONE PERSON TYPING AT A WORD PROCESSOR. THE OTHER PERSON TALKING ON THE TELEPHONE.

Figure 2–16 Our world as it really is.

SOUND: STEREO PLAYING SOFTLY. SOMEONE TYPING, SOMEONE TALK-
ING ON A TELEPHONE. BIRDS SINGING IN THE TREES OUTSIDE THE WIN-
DOW. AN OCCASIONAL CAR PASSING.

The picture and the sound. The pieces that make up our world. If we were
making a TV show out of this, to these we could add a drum roll or a crescendo to
emphasize every thought being put into words, or perhaps we could go inside our
own heads for a stream-of-consciousness approach adding a voice over saying the
words as they are typed. Or we could add the sound of a train going by or an airplane or galloping horses. Any sound we add would affect your perception of our
world. You would see it differently based on the sound that accompanied it.

If we added the visuals that accompany the sounds, such as galloping horses
and airplanes, this would affect your perception of our world even more, changing
your picture to that shown in Figure 2–17. With the addition of a sound or a visual,
we could be in the middle of the ocean or in outer space or typing away in the middle
of the jungle.

Think of how you might change *your* reality by changing the picture or the
sound. What could you add or delete to make it more exciting? Try putting yourself
in the middle of a western movie, on the moon, or in the cockpit of a jet fighter. By
doing this exercise, you will begin discovering the unlimited possibilities of life in a

Figure 2–17   Our world changed on TV by adding more sound and pictures.

television-created world. *And* you will begin realizing the fun you are about to have in creating your own TV.

Moving, Not Stills

Television is moving pictures. It is not still photographs. You can see more on television because you have this luxury of movement. Because it is moving pictures, you do not need to pose your shots like you would for a still camera. You have the ability to capture the action as it happens. That can mean that you tape while the still photographer is shooting a posed shot of the wedding party, if you like, but you will be documenting not only the posed shot on tape but also all the events leading up to it. You will have the action, not just the smiles. There's nothing more ridiculous than having everyone standing around posing for a TV camera. This is a situation to be wary of, especially for an unscripted video featuring amateurs who may be intimidated by the camera.

Think of it this way. Say you want to capture the kiss at the end of the wedding. With a photograph, you will have only the kiss. With the television camera, you could capture all the anticipation, the nervousness, the excitement of the events leading up to the kiss and what happens *after* the kiss. A subtle camera move from a wide shot to a closeup of just the bride and groom at the moment of the kiss can make this shot even more spectacular. The key idea here is naturalness. You do not want your video to look posed or staged even if every word and action in it is scripted. You want it to look real. Beware of overdirecting amateur talent, making them stiff and wooden in their movement and delivery because ultimately it is you that will come off as the amateur when the video is finished because, after all, it is your video.

On the other side, television does not move as fast as stills. That is to say, you cannot point the camera, roll tape, and stop in a split second as you can with a still camera. A television picture shot in a still photographic style will not be visually "on" long enough for your audience to "see" the shot. Think of it this way. How long do you look at a still photograph? A second isn't long enough. Is five seconds? How about ten? Maybe fifteen? Fifteen seconds is a long time in television time if you are seeing only a still picture, but if the picture is moving, maybe fifteen is not too long. You may even want to stay on it longer, depending on the action taking place.

The important thing to remember is that your television camera is not limited as is the photographic capability of a still camera. Your television camera can capture movement and can create movement with movements of its own like a pan left or a zoom in or a dolly right or a tilt up. Video pictures move.

Film vs. TV

There is a tendency to think of TV like we do a movie, so it follows that there is a tendency to shoot TV like a movie. But there is a big difference between film and video and that difference goes far beyond the kind of material on which each are

FILM

T.V.

Figure 2–18 Film vs. TV: film "sees" more in wide shots than TV does because it is shown on a larger screen.

recorded. The difference that affects you the most is where each will finally be shown. Film, for example, traditionally is shown on a big screen while television is shown on a small screen. You must adjust for this difference in your shooting style. A film can use wider shots than television. You can film a man standing on a mountain and see the entire mountain *and* the man. Put this shot on TV and you will not see the man. He will be too small in comparison to the size of the screen, as in Figure 2–18.

In TV, wide shots are used basically to establish a location or to forward the action. For example, the man on the mountain shot on TV might go like this: Tape the wide shot of the mountain to establish the location and then tape a closeup shot of the man. Edit these two shots together as shown in Figure 2–19. Now when you cut back to the wide shot of the mountain again, the viewer will know that the speck on top of the mountain that was unidentifiable before because it was so small is the man and not a tree.

Television is a closeup medium. Everything you shoot must be seen in that little box called a television set. And, as you know, that set can be as small as one inch or as large as sixty inches. Statistics say that it is most often nineteen inches in size. That is not a very big window on the television world. You must keep the size of the screen in mind at all times when you are planning your show. See the show from a TV perspective. Always ask yourself, will the shot "read" on the small screen? Will you be able to see the major action?

Focal Point

The *focal point* is that part of the picture that draws our attention and makes us watch it. Most often the major action or subject will be the desired focal point of a shot. In television you cannot be as subtle with your focal point as you can in film. The size of the TV screen is just not big enough to handle it. It is necessary to be direct, to pull the viewer's attention to the major action immediately. One of the ways to do this is with the camera's focus. If you focus on the main action, letting all other objects in the scene "fall off" in focus, this is what the viewer will watch. If the action is between two subjects, then split the focus between the two and both will hold dominance in the picture, as shown in Figure 2–20. Choose a focal point for every shot that makes the viewer watch what you want them to watch.

Angle

The camera angle dictates the way the audience perceives the show. For example, as shown in Figure 2–21, if you center your shot at waist level and shoot up, your subject becomes dominant, oppressive, or even overbearing. If you shoot down on the subject, it becomes weak, small, or insignificant. If you shoot eye to eye with your subject, you are on an equal footing. Basically, think of your camera angle as originating from your own eyes. If the shot you see in your mind looks down on the subject, then have the camera look down. If you see the shot as being on the same level

Figure 2–19 Putting the man on the mountain, TV style.

FOCUS ON FOREGROUND

SPLIT FOCUS

Figure 2–20   Focusing on one object or splitting the focus among several objects.

Shooting Down = Weak

Shooting at Eye Level = Equals

Shooting Up = Dominant

Figure 2–21 The angle at which you shoot the picture determines how the audience perceives it.

as your eyes, then shoot straight on to the subject. Make your eyes the camera lens and then angle or position the camera the way you have seen it. The point of view (POV) goes hand in hand with the angle of the shot.

Point of View

From the camera's perspective, who is seeing the shot you are recording? The answer to this question determines the point of view of that shot, how it will be seen by the audience. Is, for example, the camera making the viewer just an observer of the action or is it forcing us to look at the action through the eyes of one of the performers?

Imagine a shot of a car driving up to a hotel. Two people get out. If we shoot this from across the street, say, that is one point of view. If we shoot it from inside the car, that is another. And if we shoot it off at an angle with the branches of a bush directly in front of the camera like we are spying, that is still another. These, shown in Figure 2–22, are not only different ways to frame the shot but because of their different POV, they make the viewer feel something different about what he is watching. The shot from across the street removes the viewer from the action, making him an observer. It puts the viewer into the role of a third person looking on. The shot inside the car makes the viewer a part of the action. He has a second-person POV. Put behind the bush, the viewer becomes the observer, the person hiding. He has a first-person POV.

It is not necessary to select one POV for the entire video. You should let the shot and its intended purpose dictate the POV. For example, are the people in the car scene just checking into the hotel or are they detectives discussing a case while they drive or are they being stalked by some unknown person? Should you shoot the scene from a first-person, second-person or third-person POV *or* should you include shots from all three POVs? The question is answered by the show itself and your interpretation of that show.

Framing a Shot

What are the boundaries of the shot? When you look at the action, you see much more than your camera can capture. You must, therefore, choose the focal point of the picture and frame your shot around that focal point, keeping in mind the 3-by-4 aspect ratio of TV. The framing will determine the dominance of everything in the shot.

If there is one person standing in the foreground talking and another in the background talking, the person in the foreground will dominate. On the other hand, if you have a person in the foreground talking and another person in the background jumping off a dock, as in Figure 2–23, the background will take dominance by virtue of the action and the accompanying sound effects of splashing waters, screams, and so on. Even if the foreground person continues to talk and completely ignores the background action, the background will be dominant. That's called **UP-**

Figure 2–22   Point of view. Same scene shot from (a) **First-person POV**: camera is person hiding behind bush; (b) **Second-person POV**: inside car with principals; (c) **Third-person POV**: outside the action area, viewing scene from a distance.

Figure 2–23   Upstaging: the action in the background dominates the scene even though it is not the focal point.

STAGING, a minor activity taking over as the major activity. Upstaging could occur with just movement in the background, such as people walking by on the street, particularly if the major activity is static.

It is, therefore, important to keep this in mind when you are framing your shot. Make certain that viewers will be focused on what you want them to see, not on something else that does not add to the story.

CAN YOU TAPE WITH ONE CAMERA AND NO EDIT?

Taping with one camera *only* and *no* edit is the ultimate challenge. The logistics of doing this successfully are overwhelming and, for most, impossible. That does not mean that it cannot be done. It can, but only with the strictest of planning and attention to detail and adhering to certain taping techniques. With a one camera–no edit show, you should know where the show begins, where it is going, and where it will end before you ever roll tape in your camera. Here are some things to keep in mind on a one camera–no edit shoot.

1.   THINK AHEAD.   Burn this one into your head, because you cannot do the impossible without it. You must know at all times where you have been and where you are going. Your taping plan must be flawless.

2.   TITLES.   Record your titles *before* you start rolling tape on the show. This will give you time to get these right and thus give the show a smoother start. It will keep you from going nuts later trying to make the title fit without erasing the beginning of the show you have already recorded.

3.   MUSIC.   If you are planning a musical background for the show and your camcorder or VCR has the ability to record audio on two tracks, it would be

advisable to record the music on one of these tracks after the show is finished. You can then stretch or shorten the music to match the video. If you are going to use only the audio recorded during the taping, be sure to listen while you are recording to make sure the audio will somewhat match from one shot to another. Jumpy audio can destroy the mood and distract from the message.

4.   INSERTS.   You may want to go back and insert new pictures and sound inside the show but doing such inserts will be touchy because you will be erasing recorded material to make room for the insert. If you are not extremely careful, you may end up erasing an important part of the recorded material you want to keep *or* leaving a part that you wanted to erase. The timing must be precise to make the insert exactly fit as you want. You must not only start at the right place but you must also end at the right place. Your success in making inserts will be dependent on the capability of your equipment and your skill in using it.

5.   OVER-RECORDING.   Beware of tape rollback that will erase part of what has already been recorded. It is simply a mechanical function of the VCR or camcorder. This rollback may occur when you stop tape. Some camcorders and VCRs do not stop at the exact time you press the button. Rather, they stop, then roll back a few seconds into the recorded material. Then when you roll tape again, the tape begins recording where it stopped, *inside the already-recorded material.*

You can cope with this problem in several ways. You can make a habit of recording extra footage on the end of every shot with the idea that the rollback will erase this extra footage instead of major action. You might also use the pause button on the VCR or camcorder to avoid the problem. There is no slippage back when this is used. You cannot, however, put the equipment on indefinite pause. If it is going to be a while before you are ready to tape again, press the stop button. Then before you record again, view this last shot, pressing the pause button at the exact point where you want the new shot to begin.

Overrecording can be avoided. It just takes some thought and planning.

The one camera–no edit shoot will give you a rough but real video and it is an experience in videomaking worth having. The techniques of planning in this book will be of immeasurable assistance in helping you accept and master the challenge of the one camera–no edit shoot.

WHAT IS THE 30–3 RULE?

Finally, before we begin our video planning, let's review the **30–3 RULE.** There is an unspoken rule in television that no camera shot should last longer than 30 seconds and that no single scene should last longer than 3 minutes. The reasoning is that the audience becomes bored rapidly and that even the slowest video will appear to have more action if the shots and scenes continue to change at this rather rapid pace. The next time you watch TV put a clock on your favorite show and you will discover that

Prime Time in particular adheres strictly to this rule. You will find that these built-in multiple camera cuts and scene changes keep the action moving and fresh. Remember the 30–3 Rule as you create your video.

WHAT IS TO COME?

In the chapters to come, we will be using all the planning techniques that we have briefly introduced you to here. At the beginning of every chapter that follows, we will tell you which, if any, lists you will be using in that chapter. We suggest that you make copies of all of these NOW so that you have the forms available when you need them. You will find blank forms provided for copying purposes in the Appendix portion of this book. Make one copy of each form unless otherwise indicated. The forms to copy are

1. MASTER CHECKLIST
2. SHOW OUTLINE
3. BUDGET WORKSHEET
4. EQUIPMENT LIST
5. CREW LIST
6. FORMAT SCENARIO
7. FORMAT.   Make one copy for every scene in the video. If you do not have an idea of the number of scenes yet, make ten copies to start.
8. STORYBOARD.   Make five copies to start.
9. CAST LIST
10. LOCATION LIST
11. PROP LIST
12. SHOT LIST.   Make one copy for every location, or five to start.
13. SHOT LOG
14. EDIT LIST

You will also need

1. Clean white, unlined paper for making additional notes and making location sketches;
2. Pencils for ease in making corrections;
3. A Three-ring binder to hold all the paper relating to the show. This will become your reference book for the show.

At the conclusion of every chapter, we will give you specific instructions on what to check off your master checklist and any other instructions you may need to complete the sequence of planning covered in that chapter.

CONCLUSION

The key to a good video is a good idea, good planning, and the ability to execute the plan. Any idea can be a good one. It is the development of that idea that makes it worthy of a video. A good plan to make that video work includes: (1) choosing the subject, (2) creating the show, (3) selecting the performers, (4) planning the production details, (5) taping the show, (6) reviewing the taped material, (7) putting the show together, and (8) having the first showing. Lists that will help you in your planning include the master checklist, the outline, the budget worksheet, the equipment list, the crew list, the cast list, the location list, the prop list, the shot list, the shot log, and the edit list. The final form of the video before taping can be a format, a script or a storyboard.

It is important to always think of your video as a TV show; that is, a show that will be seen on the small screen of a television set, not on the wide screen of a film theatre. In thinking TV, you should also remember that you are creating moving pictures, not stills, and that your video will have sound as well as pictures. Consider the TV screen when you are thinking about the framing of a shot or its angle. And choose your POV, first-, second-, or third-person, based on how you perceive the story of the video.

Always remember the 30–3 rule: No shot on longer than 30 seconds, no scene on longer than 3 minutes.

Coming up next: *The Outline,* and the development of the video begins.

CHAPTER THREE

THE OUTLINE

Materials needed for this chapter:

Blank Forms

MASTER CHECKLIST
SHOW OUTLINE

Aha, I have an idea. Let's make a video out of it. We have a camera and some tape. We know how to talk TV. So, let's go.

Hold on there. Go where? Tape what? What exactly are you going to tape? Anything. Everything. Sounds like fun. I'd watch that video. Wouldn't you? Well, maybe once if someone tied you down. Of course, there is always the chance that you may get a really good shot on tape. So what! Remember this is video, not photographs. Odds are that even if you do get a really good shot of "whatever" that your audience will not stick around to see it. Television cannot be selectively viewed like photographs can be. You can scan over photographs and then concentrate on the ones that interest you, discarding the others, but in television, even if you have a fast-forward search mode on your VCR, you still have to watch everything on the tape to see the good stuff and you cannot discard any shots. So, you may find that even with the best shot, your audience may be asleep or gone long before it comes up on the TV if the rest of the video does not keep them interested.

While jumping in the car and heading for the nearest great-spot-to-shoot may work with photographs, it will not work with video. Not even you will watch a video over and over and enjoy it if it has no point. We are a television generation and we expect to be entertained when we watch TV and we really don't care whether it is your video or Dick Clark's *American Bandstand* or Aaron Spelling's *Love Boat*. We *expect* to enjoy watching. Just crunching a few great shots together will not do it for us. We have to have more.

Now that's not really as big a deal as it sounds. It only means that we must have an idea for the video and that idea must have a subject—a theme that makes the pictures make sense—that gives you a reason to watch, whether it is for pure entertainment, for information, or for the sake of nostalgia. Your video has to be about something. They make scrapbooks for bunches of pictures. They make videos to tell stories.

WHAT DO YOU WANT TO DO?

You don't have to overanalyze your video, but you do need to figure out exactly what it is you want to do.

In the last chapter we talked about some basic ideas about how to pull your thoughts together to structure and tape a great video. Now the fun begins. We start putting those ideas to work for us. From now to the conclusion of this book, we will be creating a video. We will be making our lists and schedules, formats and scripts, and complete plans for this, our first video. If you do not have an idea for a video of your own already, don't worry. We will let you borrow ours.

First of all, you need an *outline*. You should already have a copy of the outline form shown here in Figure 3–1. If not, make one now. During this chapter, we will be filling in this outline and thus organizing our thoughts about the video.

First, write in today's date. Be sure to include the year for future reference.

The Subject

What is the subject of your video? You should have a very basic answer to this question at this point, such as a wedding, a vacation, a new baby, an original rock video, a first day at school, a business meeting, a convention, a new product, or a new marketing idea. You could even be taping an original soap opera starring your friends and family. Or someone may have asked for a family recipe or asked how you started your business and rather than writing it down, you choose to put it on tape. The ideas for videos are unlimited.

If you still have not chosen a subject for a video, check out those shown in Figure 3–2 and the 101 ideas for videos listed in Chapter 14. If none of these appeal to you, we promised you could use ours and you can. We have selected one that is simple yet could, with planning and thought, show great creativity. We have chosen as our subject a wedding.

Whether you have your own idea or are borrowing ours, write in the subject of the video in the subject blank on the outline now.

SUBJECT <u>WEDDING</u>

SHOW OUTLINE

DATE: _____

SUBJECT: _____

PURPOSE: To document _____   To entertain _____

            To inform _____   To motivate _____

     To make the audience happy _____   Sad _____   Think _____

ROLES: _____   TALENT _____

         _____   TALENT _____

         _____   TALENT _____

         _____   TALENT _____

         _____   TALENT _____

         _____   TALENT _____

THE STORY

THE PLOT: _____

_____

     SUBPLOT: _____

THE CLIMAX: _____

_____

     SUBCLIMAX: _____

THE CLOSE: _____

CREDITS: yes _____   no _____

     Over close _____   black _____   show scenes _____

THE OPEN: over black _____   story _____   action sequence _____

     What does it look like? _____

THE STYLE

THE PICTURES: _____

_____

Figure 3–1 Show Outline Form.

THE MUSIC: yes _____ no _____ name it _____

   instrumental _____ lyrics _____

THE SOUND EFFECTS: yes _____ no _____ what? _____

SPECIAL ELEMENTS: slides _____ snapshots _____ film _____

   video _____ graphics/artwork _____ special effects _____

   what? _____

THE LOCATIONS: _____ DATE _____

   _____ DATE _____

   _____ DATE _____

   _____ DATE _____

   _____ DATE _____

   _____ DATE _____

THE LENGTH: less than 3 min. _____ 3 to 5 min _____

   5 to 10 min _____ 10 to 15 min _____ 15 to 20 min _____

   over 20 min (specify) _____

SCREENING DATE: _____

IF THIS IS A FAMILY OR FRIENDLY ORIENTED VIDEO, ASK YOUR PRINCIPAL
TALENT:

   WHAT IS YOUR FAVORITE . . .

      Color _____

      Song _____

      Actor/actress _____

      TV show _____

      Movie (classic) _____

      Place _____

      Other _____

   WHAT VISUALS OF THE SUBJECT CAN YOU PROVIDE?

      Snapshots Yes _____ No _____ What? _____

      Film Yes _____ No _____ What? _____

      Video Yes _____ No _____ What? _____

      Art Yes _____ No __X__ What? _____

Figure 3–1 *(cont.)*

Figure 3–2 The ideas for the subject of a videotape are endless, limited only by your imagination.

The Purpose

What is the purpose of your video? Ask yourself, "If my video was absolutely perfect, what effect would I want it to have on the audience?" This will be your purpose for making the video. You could have both a primary and a secondary purpose. For example, you may want to entertain but you may also want to inform. You might want to make your audience cry from joy or sadness or laugh hysterically. You may just be documenting a family event. Whatever it is, you have a reason for making this video. What is it? This reason is your purpose and what you should check on the outline.

For our wedding video, our purpose is to both document the event and to make it memorable and fun, involving our audience enough to make them a part of the video and, therefore, to get them to react emotionally. Quite simply, we want to make them cry with happiness at the beauty of not only the video but of the subject. We would, therefore, check on our outline:

PURPOSE: To document   __X__   To entertain _____

          To inform _____   To motivate _____

          To make the audience Happy __X__   Sad _____

          Think _____

The Performers

Who will be in your video? What roles will they play? There are different kinds of roles in every video. There are the major ones called principals. There are the secondary ones called supporting principals. There may be an announcer. And there are all those other folks who will appear in the video but really have nothing specific to say who we will call the extras. At this early stage in the development of your video, you should have a basic idea of what roles will be available in your video. You may also have at least a beginning idea of who you want to play those roles. It is even possible that the video has "natural" performers for certain roles. Our wedding, for example, has certain natural roles and performers, such as the bride, Mary, and the groom, John. You might argue that a video like this has all natural roles and performers. Not true. We could create special roles based on our format for the video. In any case, even with a video that has all natural roles and performers, the way you decide to tell the story will determine who is cast as a principal and who appears only as an extra.

If you are doing a video with no natural performers, you will be casting all the roles. While you may not now know who will play all the roles in your video, you should have an idea of what the roles will be. So let's start there.

For the rock video, you might write in, for example, **LEAD SINGER, BAND, SEXY LADY,** and **ZOMBIE.** For our wedding video, beyond the obvious roles, we might add **BEST FRIEND, GRANDPARENTS,** or **LITTLE BROTHER** with the idea that we have a special role in mind for these. Our cast list would look like this:

ROLE BRIDE _____   TALENT _____

ROLE GROOM _____   TALENT _____

ROLE BEST FRIEND _____   TALENT _____

ROLE LITTLE BROTHER _____   TALENT _____

ROLE GRANDFATHER OF BRIDE _____   TALENT _____

You may add more roles as the video develops. Generally, you will not want to list extras on this beginning cast list. If you later decide to give extras credit for their appearance in the video, you can get their names and add them to your completed cast list.

Now that you have an idea of the roles for the video, you can begin to cast. At this time, write in only those people you are fairly sure will be playing certain roles. You may want to recast later after you have had a chance to audition all your talent. In the case of our wedding video, we have certain natural performers, so we would write in the names of our naturals at this time. If you have natural roles and performers, the cast list is relatively easy to fill in except for any special roles. If you have no naturals, it requires more thought. For now, fill in on your outline those roles and performers that you can. We will be adding to these later as we continue the development of the video.

WHAT IS THE STORY YOU WANT TO TELL?

Every video has a beginning, a middle, and an end, as shown in Figure 3–3. So will yours. It should also tell a story whether it is a fictional drama, a marketing plan, or the documentation of a wedding.

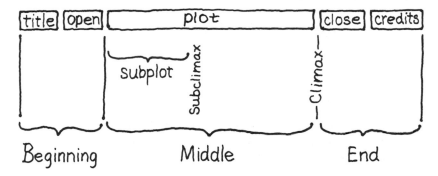

Figure 3–3   Every story has a beginning (title and open), a middle (plot, subplot) and an end (climax, subclimax, close and credits).

Our wedding video, for example, is more than just a wedding. It is John and Mary's story and the way we tell that story will forever affect the way we see them. There are lots of ways we could tell their story. Told one way, their wedding is the end of one relationship and the beginning of another. Another way, the wedding is the end to a great love story with the classic *and they lived happily ever after* conclusion. Or we could tell it even a different way with the story being the wedding itself, from the initial preparations up to the *I do*. Or still another approach would be to tell it as the story of two different people; their lives and how those two lives become one, or in still another, we could create a video with the wedding seen through the eyes of the guests with each offering a unique point of view.

Whatever the subject of the video, you will have more than one way to tell the story behind it. Someone will have to choose *one*. You cannot do all of them. Choose one approach and stick to it.

How would you tell the story of the wedding? Before we make a choice, let's break the video into pieces so we can better determine exactly what the story is as we see it.

The Plot or Theme

What is the core of the video? What is the central action? What will the bulk of the video be about? What is the main story line? This is the *main plot*. In our case, let's say that the main plot is the wedding itself. We are going to build our video around that particular action. Our goal then is to build a story around the wedding that will make it interesting and exciting and thus create a video that people will want to watch over and over. We would, then, write on our outline:

PLOT   THE WEDDING

What is your Main Plot? If your subject were *A Special Vacation,* you might answer that your main plot was:

PLOT   FUN ABOARD THE CRUISE SHIP

or

PLOT   HIKING IN YOSEMITE

In addition to the main plot, you will most likely have a subplot or, in other words, a smaller story line that works with and ties into the main story line. For example, to your main plot of **Fun Aboard the Cruise Ship,** your subplot might be:

SUBPLOT   EATING

or to your **Hiking in Yosemite,** your subplot might be:

SUBPLOT   COOKING OVER A CAMPFIRE

The main plot and the subplot are intertwined to make up the complete story. One adds to the other and at the same time, one supplies variety and relief from the other. In other words, you may have more than one story going at the same time. This way the audience is less likely to get bored. It is even possible to have more than two stories going at once. You could have many, adding even more to the complexity of the video. For this exercise, let's try to keep it relatively simple with only one subplot and one main plot.

On our outline for our wedding video, we would write in:

SUBPLOT <u>JOHN AND MARY'S COURTSHIP</u>

As you should begin to see, our wedding video is going to be more than just the wedding. It is going to build from John and Mary's first meeting to their eventual wedding. The wedding is the main plot and, therefore, the other story will intertwine with and expand on that story.

The Climax

When you are telling a story, you are putting together a series of events that ultimately lead to a final, culminating event. This event is the highest point of interest or excitement and leads to a decisive turning point in the story. This is called the *climax*. What you perceive as your climax will affect the way you tell your story and thus the way that the audience sees the story. In a sense, the climax is the goal of your main plot. If our main plot were the courtship and our goal was to get John to ask Mary to marry him or vice versa, then the proposal would be the climax. On the other hand, if the goal was to get John and Mary married, then the *I do* of the ceremony is the climax. The second is true for our wedding video so we have written on our *outline:*

CLIMAX <u>I DO</u>

If you have chosen a different subject, what is the climax of your video? If you were taping the Cruise Vacation, it might be:

CLIMAX <u>THE LAST NIGHT OUT</u>

or

CLIMAX <u>GOING ABOARD THE SHIP</u>

You can see that you will have a different video depending on which one of these you choose. One video would be the story of the cruise itself with the last night out being the conclusion. The other would be the story of how you finally achieved your dream of taking a cruise, and, essentially, the video would end with you going aboard the ship.

You could say that the climax is the end of the story. There are things that happen after the climax but, for the most part, they are only things that tie up loose ends. The action or story does not, as a general rule, progress beyond the climax. It is, of course, possible to have more than one climax. If you have a main plot and a subplot, you will have a climax and a subclimax. Each story you begin must have an end. Sometimes, they will end together with one climax becoming the climax for all the plots. Sometimes not. It is important to think through each plot to its logical conclusion or climax.

For our wedding video, we have two climaxes. The subplot ends with the proposal while building the major plot toward the wedding ceremony. If we were doing the cruise vacation and our main plot were **Fun Aboard Ship** and our subplot was **Eating Aboard Ship,** our main plot would climax with the **Last Night Aboard Ship.** It is conceivable that our **Eating** subplot could end that night also with a spectacular meal or we could choose to end that story line the preceding night at a Midnight Champagne Buffet.

You should begin to understand how one story can interweave with another and how one story can end while another goes on. So now, on your outline, write down what you believe to be the climax for each of your plots. For our wedding video, we would write:

CLIMAX "I DO"

SUBPLOT CLIMAX PROPOSAL

Our plan is to get John and Mary to reenact the proposal so that we can make it a part of the video and to end the video with the *I Do* with the reception a part of the closing sequence.

The Close

The close is everything that happens after the climax and usually includes the credits. We will, however, deal with the credits separately.

With the close, you wrap up all the loose ends and end the video. For example, with our wedding video, our close will be a series of shots including:

CLOSE RECEPTION AND LEAVING FOR HONEYMOON

And, if we can get the footage in time to make it a part of the video, we might include **John and Mary in their new home.** We would include this "at home" video to reinforce our romantic theme, *they lived happily ever after*.

Your close is a key part of the video. It may be what the audience will remember most because they saw it last. You want to make it strong and a reinforcement of the emotional reaction you set out to get from your audience. Ideally, when the tape stops, your audience should be reacting just as you wanted them to react.

The Credits

The credits are at the very end of the video and are simply a list of the performers, producer, director, camera operators, and the entire production team. Credits are

THE CLOSE

BLACK

SHOW SCENES

Figure 3–4 Some of the ways credits can be added to the show. Speed filter courtesy COKIN® Filters.

not absolutely essential to your video but they are a courtesy *thank you* to your helpers and a professional touch that will give the production more credibility with the audience.

Three common ways to add credits, as seen in Figure 3–4, are to put them over:

1. The close
2. Black
3. A sequence of actions that happened inside the video.

You can even combine these different ways of adding credits. For example, the credits could begin over the close and continue as the close ends and the video goes to black with the credits ending over black only.

In our wedding video, we would choose to do it just like that with the credits supered over our shots of John and Mary in their new life, the video finally going to black and the credits continuing over this black. Of course, we would continue the romantic music to the final credit to maintain the mood of the video. We would, therefore, write on our outline:

CREDITS: Yes ___X___   No _____

Over close ___X___   Black ___X___   Show scenes _____

What would you do if, for example, your video were the cruise vacation? Would you roll your credits over black or would you put them over video of you departing the ship or perhaps super them over a montage of shots of you during the cruise? You should choose the one that best accomplishes your purpose and keeps your audience's attention right to final video, keeping in mind, of course, the capability of your equipment.

The Opening

We have saved this for last because, while it is the first thing your audience will see, it is a product of the completed video. It sets the mood for what is to follow. It is the hook that will keep them watching. It should immediately tell the audience what kind of video they are about to see, such as an action-packed video, a romantic video, an informative video, or one just for fun. The opening scenes are extremely important because this is the first thing the audience sees, leading them to make an immediate judgment decision about the video, your own abilities, and whether they want to watch. A poor open can doom even a great video because your audience could mentally check out before the good part even comes on. There are three basic kinds of openings, as seen in Figure 3–5.

An open may be no more than the name of the video over black or it can be what is called a tease, a sequence of actions that gets the audience's attention and makes them want to see more. The sequence of actions can be the actual beginning

TITLE ONLY

STORY WITH TITLES

ACTION SEQUENCES WITH TITLES

Figure 3–5 The different ways the title (and subtitle if you have one) can be put on a show.

of the story you are going to tell, say a murder happening, or they can be a sequence of shots that will be seen within the video, say, the pursuit of the murderer. In this last case, you would, of course, choose shots that are exciting to capture the audience's attention and thus, in a way, your opening becomes a "commercial" of what is to come. Either of these types of openings can be extremely effective. On the other hand, with the right music and maybe a voice over to back up the titles, you may not need pictures at all to capture the audience. Think of the *Star Wars* opening. It was just words and music but it certainly did get your attention.

But the choice is yours. Would you choose to begin your video with:

1. Title over black
2. Title over the beginning story
3. Title over action sequences taken from inside the video.

In the case of our wedding video, we chose:

THE OPENING: Over black _____   Story _____

            Action _____   Show sequences   X

Our plan at this time—and we may change this as we progress—is to open with shots of John and Mary together, falling in love, with the appropriate romantic music in the background and titles over. With this opening, we will set the mood for the video. Our purpose for making the video, as you will recall, is to make our audience "cry with joy" and we will begin with this opening romantic sequence, creating our own music video under titles.

As you can see, the open is not just something that is at the beginning of the video. It is a part of the total effect you are trying to create. It should from first video begin to accomplish your purpose in making the video. In some ways, the open may even be more important than the bulk of the video, since it is that first impression that will keep or lose your audience. How many times have you watched a TV show and turned the channel or just got up and walked out when the open turned you off? Think about it. And then choose your open with care.

HOW DO YOU TELL THE STORY?

In television the most important ingredient to a good show is your ability to see and hear it before you ever roll tape. If you have made good plans, this will not be difficult, because piece by piece you will put it all together. And if by some chance, you still cannot "see" the show, you can make basic storyboards. We will get into that later but for now suffice it to say that visualization is number one on your list. Some people can do this easily. Others need to use some tricks of the trade like the storyboard. Either way, what you see and hear and how you see and hear it will create your style and the mood of the video.

Style is a combination of how you see the video and how you go about transferring that visualization to video with pictures, music, sound, and special effects, as demonstrated in Figure 3–6. Style is a very individual thing and you may develop a certain style that marks all your videos as yours. Michael Mann, for example, has a certain style using dramatic and colored lighting that can be seen in both *Miami Vice* and *Crime Story*. Aaron Spelling has another type of style that utilizes formula storytelling with tight framing and realistic locations evident in *The Love Boat* and *Hotel*. And you may see a similarity between the styles of *Magnum P.I.* and *Rockford* because both were produced and created by Stephen J. Cannell.

Figure 3–6   Style is a combination of how you see the video and how you transfer that visualization into what you record on tape.

Style is created by the lighting you use, the story line you follow, the camera work, the music you use as background, the casting, all of these as they come together in the video. Style is what you "see and hear;" your concept of the video. It is, after all, *your* video even when it is about someone else. Every video has its own individual style, and while that style is representative of your own, it is unique to that video as well.

The Pictures You See

Style is the pictures you see (Figure 3–7). You may see action in everything. A wedding can be fast with frantic activity. On the other hand, you may see romance in everything and play the wedding soft and sweet. Your perception of the subject will directly affect the pictures you see in your own head. At this early stage in the development of this video, you want to ask yourself what kinds of pictures you see for this video. Are you going to concentrate on the action of putting the wedding together or are you going to concentrate on the love between John and Mary? Do you

Figure 3–7   Style = pictures.

see close intimate shots of the couple and family and friends or do you see crowds and ceremony? Or do you see a combination of both? For our part, we see:

PICTURES   CLOSEUPS: CREATE AN INTIMATE RELATIONSHIP BETWEEN THE COUPLE AND THEIR FAMILY AND FRIENDS

The Music You Hear

Style is the music you hear (Figure 3–8). Music can make a video. Music can add meaning to an otherwise meaningless picture. Music can add emphasis to a situation or action. Music has its own way of creating pictures in the mind without your actually having to see them.

Figure 3–8   Style = music.

Knowing this, you must choose your music carefully and use it wisely. To put a video together and just run any old music under it will, as a rule, not work. The audience will just get bored with the music and this will more than likely make them equally bored with the pictures. You need a balance between pictures and music so that they complement each other. Music can add to or take away from what you are trying to do. And remember, music comes with and without lyrics and that leaves you with still another choice to make.

Keeping all this in mind, we want to select for our wedding video a theme song that will be appropriate and help us create the mood. We may add a second song for the close or elsewhere in the video, but for now, we would write on our outline the song we are hearing as we are visualizing the video:

MUSIC: Yes ___X___   No _____   Name it _their favorite love song_

Instrumental __X__   Lyrics _____

The Sounds You Hear

Style is the sound you hear (Figure 3–9). Sound can be critical. Think of watching a car chase without the sound. With no screeching of brakes, skidding of tires, car acceleration, and so on, it would be dull and lifeless, certainly not realistic. For our purpose, *sound effects* are those sounds we add to what already exists. If, for example, we added the roar of a jet engine to the car in the chase, the car becomes more

Figure 3–9   Style = sound.

powerful and the car chase becomes more intense. In this case, the jet would be the sound effect. Sounds that are recorded as they happen, such as crowd noises, are not considered sound effects. Sounds you add in the edit are sound effects. For example, if you shot a crowd sequence without a microphone and you wanted to add the sound of the crowd later in the edit, you would be adding a sound effect.

At this point in the development of the wedding video, we can think of no sound effect that we will want to add. We do, however, as always, reserve the right to change our minds later. For now, we would mark our outline as follows:

SOUND EFFECTS: Yes _____   No   **X**   What?_____

The Special Visual Elements

Style is those special elements you add (Figure 3–10). Now you can begin to think about those special little touches that will make the video exciting and that will make it yours. You should learn to think beyond the action you are planning to tape. There are all kinds of other visuals outside what you tape that can be used in your video to make it better. When considering what other visuals are available to you, remember you are creating a special video for a limited audience. You will not be charging your audience to see it and you will not be distributing it to anyone beyond a very select list of family, friends, or colleagues. These facts give you some access to the works of others. Certainly if the video is for the wedding party only, you can use that special song you love so much or add those special effects from your favor-

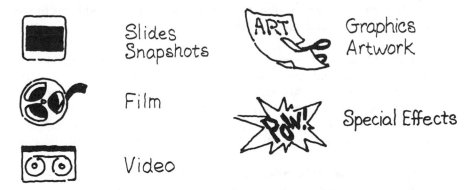

Slides
Snapshots

Graphics
Artwork

Film

Special Effects

Video

Figure 3–10   Style = special elements.

ite television commercial or include the couple's favorite television performer in the reception party. If, however, you are creating a video for business use, then you should get copyright clearance before using other people's material. The important thing to remember is that you are not limited by what you can tape yourself or by what photographs or special effects or graphics you can create yourself. Keep that in mind and let your imagination run wild with the possibilities as you consider the following sources.

Slides and Snapshots.   Just because a visual element did not originate on video does not mean that it cannot become an effective part of your video. Old or new, slides and snapshots can add color and charm. You could use them in a visual montage effect to go from the past to the present. We might follow our groom, for example, from his first baby picture through his preschool years right up to high school and his first meeting with Mary. Or we could put our groom with a gorgeous blonde in a sports car simply by taking a photograph of him and pasting that over a photo of the girl and the sports car. Crude but perhaps cute if you keep your editing tight, not leaving this visual up longer than a few seconds. Just long enough for the audience to get the effect, but not long enough for them to criticize the crudeness of the effect.

Ask yourself what pictures are available to you. Then ask yourself what others you might want to use. And finally, what can you do to those pictures to make them more effective? Pictures can be a really effective tool in creating a mood or in creating your style.

With regard to slides, you should not be afraid of them just because they are slides. You can easily have them transferred to video at your local video store, or you can forget the fancy transfer process and do it yourself. Just project the slide onto a screen and tape from there, or you can have the slide made into a print and shoot the print. You can even create a kind of overlapping dissolve effect from one slide to another by using two slide projectors.

Film.   Before videos, there were home movies and it is likely that somewhere there exists a piece of film in someone's attic that will add enormously to the telling of your story. Ask around and check your own attic.

Other Video.   Again, there may be shots already on tape that will add to your video. Ask around and preview all possible material. Also, here is where your stock footage comes in. All those other shots you have on videotape stored away somewhere or that are already a part of another video. You can reuse them or use them for the first time. For example, maybe now you can use that gorgeous sunset you put on tape because you just could not let it pass into oblivion.

Graphics or Artwork.   This may be hand-drawn pictures or computer-generated art or a montage of works of art that you found in a magazine, such as a picture of Mary's wedding dress as seen in the magazine, cut to a shot of Mary wearing the dress. This may also include words or titles for effect, like the word *WOW!* supered over the picture when Mary meets John for the first time.

Special Effects.   Sometimes special effects add to a video and sometimes they distract from it. You should be careful when you are planning these to make sure that they do the former and not the latter for your video. Effects can be created simply by hand or they can be generated by your equipment. Those you can generate with the equipment will be limited by the capabilities of your equipment. Those you can create by hand will be limited only by your imagination.

For example, let's say that you want a shot of the night sky full of stars. It is virtually impossible to shoot the stars. Even if you could find the perfect clear night, your camera still would not see what your eyes see. It would see only black. So how do you tape stars? With fancy equipment, you could do it electronically, creating them with a computer effect. If you do not have that, you could do this:

Get a piece of black poster board. The size is immaterial as long as it is TV perspective, meaning that it should conform to the 3 by 4 aspect ratio of your television screen and not be square or round. Now, punch tiny holes in this poster board, punching from the black side back. One punch equals one star. Now put the black poster board—now your "star field"—in front of a bright light. Position your camera, adjust your light, add more stars if you want, and roll tape. What you will have on tape will be the night sky. If you want the stars to twinkle, just move the light around while you are taping.

Simple, crude, but effective. Imagination. That is the key to creating great special effects.

We have chosen for our wedding:

SPECIAL ELEMENTS:

| Slides _____ | Snapshots X | Film X | Video X |

Graphics/artwork _____   Special effects _____

With these elements we can create a wedding video that not only documents a special moment for John and Mary, but also a video that shows some imagination in the use of visuals or sounds and music from their own background.

You can see that there are a lot more things than just video to consider and that your choices of what you want to use are limited only by your own thinking process. We encourage you to let your imagination run wild, to think of anything and everything. Write it down on your outline. No doubt you will think of even more elements as you continue the planning process. Write those ideas down, too. Then you have something to work from, a place to begin. Whether you actually do everything will be determined ultimately by whether it works with the final version of the video and whether or not you have access to the elements and equipment needed to do it.

WHERE CAN YOU TAPE YOUR SHOW?

A television show is either taped inside, called *interiors,* or outside, called *exteriors.* In addition, it is either taped (1) *in the studio*—your designated home base—in the business, a professional warehouse arrangement that is equipped to supply your every need; or (2) *in the field,* anywhere away from the studio. If you are in the field, you are said to be *on remote* or *on location.* If you are in the studio, what you are able to shoot is determined by the size of your surroundings and by your set-construction capabilities. If, however, you are in the field, you have unlimited choices of what your surroundings will be. You can select a private home or a public place like an amusement park or special attraction. If you were taping for broadcast television, you might have to get all kinds of special permissions and pay a fee to tape at a public place. But since you are not taping for broadcast, you can go virtually anywhere, and even if you plan to bring talent along, you should have relatively easy access to all kinds of public places. You could tape on the Golden Gate Bridge, on Miami Beach, or at the top of the Empire State Building. The important thing to remember is that you have all kinds of locations open to you and any that are relevant to the video are worth checking out to see if you can tape there. It never hurts to ask if you have doubts about whether you need special permission to tape somewhere.

Locations are important to the video. The wrong location will make a bad video. A good location can make a good one. You do want to remember to select your locations based on the story and the style of the video and not get carried away by the possibilities open to you. It is a good idea to make a list of what you believe to be good locations for the video and then to check these out with your camera before you lock into any one of them. There are actually some locations that look terrific to the eye but rather bland to the camera. The camera, as you will remember, is limited by what it can see. It cannot see as wide as you can and it cannot capture all those sounds, smells, and vibes that might make a particular spot just perfect for the eye. What you see can be deceiving when it comes to putting it on videotape.

A test shot with your camera will allow you to look at the framing, see the lighting conditions and hear all the ambient sound of the location, and help you decide whether it is a good location for your video.

The locations you select will help to tell or distract from the telling of your story. If, for example, we taped our wedding totally in the church and never went beyond that, we would have a very narrow story and a fairly uninteresting one at that, since most everyone has been to a wedding before. There are few surprises in that story alone.

In thinking about our wedding video, we begin our list of locations with those chosen by the bride and groom:

Locations

CHURCH
RECEPTION HALL

These we *must* cover. Beyond that, the locations become a factor of the way we perceive the video. We have already decided in the development of our outline that we want to cover John and Mary's courtship and nostalgically deal with their childhood sweethearts. In addition, we thought it might be fun to have Mary's little brother give us a tour of her room and to interview Mary's closest girlfriend. We also want to be there with John and Mary immediately prior to the wedding to capture the jitters and finally, we want to watch them leave on their honeymoon. With all this in mind, we have chosen these additional locations:

Locations

MARY'S HOUSE
THE PARK

At this time, our plan is to hopefully tape a dynamite sunset in the park that we will use as the close for the credit roll.

We may also use stock footage, still photographs, film, or other video in the show but we do not need to plan to shoot these on location since they already exist.

Now that we know our locations, we can break them down into interiors and exteriors at this time or we can wait until after the format and/or script are complete. At this point, we have open to us the option of taping inside or outside in every location we have chosen. Let's leave it at that.

WHAT ARE THE INITIAL SPECIFICS AND DEADLINES?

How long is your video going to be? When will you tape? When will the video be ready to view? Answering these questions now will save you some time and trouble later and give you certain deadlines that will push you into completing the video. The biggest problem with most creative endeavors is the time factor. If you do not have deadlines, you could work on it forever trying to make it perfect. While this process eats up time, it will also more than likely make the video longer and longer. And long does not usually translate into *good*. It is more likely to translate into *boring*. Deadlines for taping and screening and a predetermination of length for the video give you a starting and a stopping point, but don't try to rush the process. Insist on a realistic time frame for the completion of the video.

Taping Date

First, what will be the main taping dates? You may have the bulk of your taping time dictated by the occasion, like our wedding. However, you may want to schedule other taping as well, like special interviews or catching shots of Mary trying on wedding gowns. Any taping dates that you plan now may be changed later or added to as the video develops, but at least you will have the beginnings of a taping schedule by writing on the outline next to the location any taping dates that you now know about.

Location

CHURCH   November 4
RECEPTION   November 4
MARY'S HOUSE   November 1, 2, or 3
PARK   November 1, 2, or 3

Length

How long is the video? You can make it any length you want, but a very long video could turn your audience off even if it is a great one. A too-short video could do the same. As a guideline, consider the following:

1.   MUSIC VIDEOS.   These are usually 3 to 5 minutes in length.
2.   BROADCAST TELEVISION.   All half-hour shows on broadcast television are 22 minutes and 38 seconds long. An hour show is approximately twice this length. The remainder of the half-hour or hour is devoted to commercials.
3.   CORPORATE TV.   Most big corporations make videos for their employees or customers. Time and experience has taught them that the magic number is 10 to 25 minutes for a video. After that, the audience gets bored.

With all this information, how long do you want your video to be? The wedding of John and Mary will actually take 3 to 4 hours with the reception, but we certainly don't want the video to run that long. A well planned 15 to 20 minutes of show would be a lot more exciting and be watched a lot more often than a 3-hour extravaganza shot in real time.

Once you decide how long your video will be, you can shorten or expand it as you go along but at least you will have a goal to shoot for. Indicate the proposed length of your video on the outline. For our wedding:

THE LENGTH:

less than 3 min _____   3 to 5 min _____   5 to 10 min _____

10 to 15 min _____   15 to 20 min _X_   over 20 min (specify) _____

Premiere Showing

When is the world going to get to see your finished extravaganza? When you are scheduling this special occasion, be sure to give yourself enough time to put everything together to make it as great as it can be. Don't tape on Monday and plan to show it Wednesday night, for example. The time you spend after the taping is just as important and maybe even more important than the taping itself.

There is a saying in the business, "we'll fix it in post." You will find yourself doing the same thing, taking a sequence that doesn't quite work and making it work with clever editing. Or you could find yourself eliminating shots you thought were important but for one reason or another just do not work. A video is more than just rolling tape. It is a total experience from beginning idea to premiere, and postproduction is one of the major aspects. In short, you need time to review what you have recorded and to make decisions about your video. You need time to step away from it for a while so that you will have a fresh, more objective perspective. Set a screening date that will allow you this important time.

Again, as the video develops you may change your mind about the screening date, but at least you will have a date you can talk about to those performers and potential viewers who will be waiting anxiously to see your video. Write the screening date on your outline.

SCREENING DATE <u>December 15</u>

ANSWERING ONE LAST QUESTION

One last thing before we move on. If the subject of this video is a family or friendly event, not a business function or fictional situation such as a soap opera, ask your principal talent the following questions.

WHAT IS YOUR FAVORITE . . .
Color? Song? Actor or Actress? TV show? Movie? Place?

And then . . .

WHAT VISUALS OF THE SUBJECT CAN YOU PROVIDE?
Snapshots? Film? Video? Art?

This information will help you create a video that represents the people who are your principals. Our wedding video, for example, can benefit from knowing the answers Mary and John would have to these questions and using those special snapshots or video they took of themselves. These elements could make the video. If it is appropriate for your video, be sure to ask these important questions.

CONCLUSION

Now, having completed our OUTLINE, we can begin to "see" the video. We have now determined our subject, isolated our purpose for doing the video, and done some preliminary casting. We have isolated our plot, determined our climax and decided on the story line we will pursue. We have made some basic decisions about credits, the opening, and the style we will attempt to create with pictures, music, sound, and special effects. We have chosen some locations, decided on a length for the video, and began our scheduling process by deciding on a premiere screen date. And lastly, if this is a spontaneous video, we have asked our principal talent some important questions about their likes and dislikes.

We will be using this outline as the base for our checklists and planning. Preparing this outline has essentially been an exercise to clear our minds and direct our attentions and ideas to the video we are considering producing.

Before we make the final production commitment, we need to examine one more thing, the *cost*. We will do this in the next chapter.

Coming up next: *The Budget*.

TO DO

1. Complete your Outline.
2. Check it off the Master Checklist.

CHAPTER FOUR

THE BUDGET

Materials needed for this chapter:

Still Working On

MASTER CHECKLIST

Blank Forms

WORKING BUDGET

This is the shortest chapter in this book, but the most important, because its subject is money. Namely, how much is your video going to cost? Nothing, you say. Not true. Even if you tape a family gathering for your own personal use, it costs something to do it. It may be no more than the cost of the tape you use in the camera and your own time in doing it, but it *will* cost something. Who is going to pay for that cost? You? Is the tape for your own use? Are you doing it for a friend? Or are you making it as a service or an assignment for a business? Is it fair for you to bear the full burden of the cost of the video? Should someone else share that cost with you? And should you be compensated for your time?

Our example throughout this book has been a wedding video. We are doing this video for John and Mary and Mary's parents are paying the cost of producing the video. As is the custom, we will discuss the budget for the video with them immediately after we prepare it and certainly before we take the production any further than the outline stage. Once they approve the budget, we will continue with the development and taping of the production. This is the procedure you should follow with your videos.

While you may begin making videos for your own use, as your skills in making them become better, your family, friends, and colleagues will begin to look at you as their own personal television producer. Inevitably they will begin to ask you to roll tape on something for them. It is then that you will want to be able to prepare a budget, submit it for their approval, and get an advance on the cost so you can begin your planning. It would not be fair for them to expect you to expend your own money on tape, props, transportation, or whatever in order to create their video. They may expect you to give them your time for free. If you do, that is more than enough for free. You may even decide that that is too much and charge them for your time as well. Your expertise is, after all, worth something or they would not be asking you to do the video in the first place.

Creating a budget is easy. Sticking to it may be hard. But let's deal with the easy part for now. First, let's begin working on our budget using the worksheet shown in Figure 4–1, which you have already copied for this purpose.

Once we have completed this budget, we will have an estimated cost which we can then present to the person paying the bills, who in the business is called the executive producer. Then throughout the production we will keep receipts of all expenses. When we deliver the tape, we will deliver also a final bill less the advance made for expenses.

One word about an executive producer. If someone else is paying the bill, then that person traditionally has final approval of the video. If the executive producer wants changes, even if they are requested after the video is totally edited, assuming he is prepared to pay for any additional costs, it is your job as the producer to make these changes. Sometimes in the business a producer can request total creative control and get it. Then if the executive producer does not like the video for some reason, that is just tough. If you feel the need, you might want to negotiate this creative control, but do it before you begin. After the video is in production will be too late.

WHERE DO YOU BEGIN?

When calculating your costs, begin at the beginning.

Videotape.   What is the tape going to cost? How much do you plan to use and of what quality? All tape does not cost the same. The higher the quality, the higher the cost. Generally, it is a good idea to make the initial recording on the highest quality tape you can afford to buy. That way you are assured of making an original recording that is as good technically as your camera can produce.

Location Fee.   Will you have to pay a location fee? Generally, the answer to this question will be no, but on occasion you may be asked to pay a nominal fee. Ask about your location if you are not sure.

Show Title _____

Date _____

BUDGET WORKSHEET

Note: Estimate all costs. It is not necessary to make any final decisions about locations, equipment, crew, edit, or set at this time. This budget is a best-guess estimate of all expenses for the proposed video.

VIDEOTAPE

    Cost per tape $_____

    No. Cassettes   ×_____

                                  Total $_____

LOCATION FEE

    (Place, if you know)

    _____ $_____

    _____   _____

    _____   _____

                                    Total $_____

TRANSPORTATION

    Gas $_____

    Parking   _____

    Toll charges   _____

    Airline tickets   _____

    Misc.   _____

                                    Total $_____

PROPS

    Buy $_____

    Rent $_____

                                    Total $_____

Figure 4–1 Budget Worksheet form.

EQUIPMENT

(Name it, if possible)

Video _____

_____

_____ $_____

Audio _____

_____

_____   _____

Lighting _____

_____

_____   _____

Misc. _____

_____

Total $_____

CREW

Cost per $_____

No. crew members   ×_____

Subtotal   _____

Plus other   +_____

Total $_____

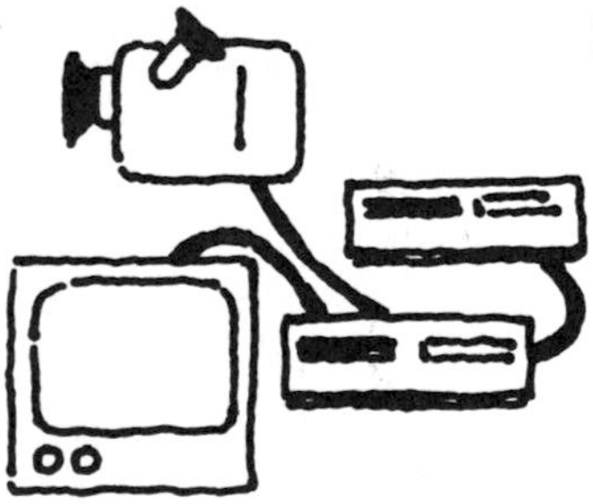

EDIT

Rent/buy equip _____

_____ $_____

Subtotal $_____

Postproduction facility

cost per hour $_____

No. hours est.   ×_____

Subtotal $_____

Total $_____

Figure 4–1 (*cont.*)

TELEPHONE

    Estimated cost $_____ $_____

ARTWORK/GRAPHICS

    Misc. expenses $_____

    Artist   _____

                       Total $_____

CATERING

    Estimated cost $_____ $_____

MUSIC

    Buy records/tapes $_____

    Original music $_____

                       Total $_____

ADDITIONAL VISUALS

    Photos $_____

    Video   _____

    Film   _____

    Other   _____

                       Total $_____

POSTAGE/DELIVERY SERVICE

    Stamps $_____

    Delivery $_____

                       Total $_____

PHOTOCOPYING

    Estimated cost $_____

    Supplies   _____

                       Total $_____

SET

    Planning $_____

    Construction   _____

                       Total $_____

Figure 4–1 (*cont.*)

CAST

    Expenses $_____

    Per diem   _____

                         Total $_____

DUB COST

    Cost per tape $_____

    No. cassettes   ×_____

    Subtotal $_____

    Dubbing fee $_____

    No. cassettes   ×_____

    Subtotal $_____

                         Total $_____

WARDROBE

    Buy $_____

    Rent   _____

                         Total $_____

MAKEUP/HAIR

    Supplies $_____

    Misc. expense $_____

                         Total $_____

TOTAL ALL $_____

MISC. EXPENSES

    Total all $_____

    Add 10%   ×_____.10

                         Total $_____

YOUR FEE $_____

TOTAL BUDGET $_____

Figure 4–1 (*cont.*)

Transportation.   Will you have any transportation costs? Gasoline, incidental parking, bridge or toll-road charges, bus, train, or airplane tickets, and so on all figure in here.

Props.   Will you have to purchase or rent certain props needed for the video? A ring from the dime store for a "cute" sequence in the wedding video or a ring coming out of a cereal box. Anything that can be picked up and carried off easily, not furniture, is a prop.

Equipment.   Do you have enough video equipment of your own? If not, can you borrow enough? Will you have to rent anything additional? You should specify exactly what you expect this rented equipment to be. Lighting fixtures, another camera, a VCR, whatever.

Crew.   Do you have enough volunteers to run the equipment? Are you going to pay them at all? It is always nice to offer some token payment for their time whether it is money or a gift. Perhaps in the wedding, the bride and groom could treat your crew like a part of the wedding party and give them a small gift just as they do the other members of the wedding. Or perhaps a twenty dollar bill would be easier and more welcome. This is strictly up to the executive producer, since the dollars are coming out of his pocket.

Edit.   Will you have enough equipment to edit the show properly? Will you need to rent additional equipment? Will you need to go to a local VHS/BETA/8MM edit house to edit? If you do need to rent edit time at a professional edit house, find out what the hourly cost is and multiply this by how many hours you think you will need to edit to determine a cost. It is advisable to increase the edit time by at least one half beyond what you think it will be. Editing always takes more time than you plan. Mostly because once you are there, you keep finding new and exciting ways to do things and there is a tendency to change your mind over and over. We are going to try to eliminate most of this with planning but inevitably there will be some of this creativity after tape rolls.

Telephone.   You may be required to make some telephone calls that cost you toll or long-distance charges. If you are not footing the bill for the video, it is proper for you to include these on the budget. Include an estimate in the budget so this will not be a surprise later.

Artwork/Graphics.   Are you going to have some pictures drawn for the show? Will you need to prepare title cards or end credits in a hard art form? You may be able to do most of your words in the edit but pictures or graphics may have to be drawn by a professional artist. Include here also any expenses you may have for poster board, press type, pencils, and the like.

Catering.   This is not as unusual a cost for a home production as you might think. The executive producer may elect to feed the crew on the rehearsal day or the shoot day. Maybe it is nothing more than coffee and donuts. Whatever the cost, it should be added to the budget.

Music.   Will you need to purchase a special record, cassette or disc for the show? Or will you be hiring a band or a singing group?

Additional Visuals.   Will you need to copy a photograph to videotape or have it printed larger? Will you need to acquire some special visual the executive producer wants? Will you need to have some film transferred to video?

Postage/Delivery Service.   Will you be spending money on mailing letters or the dubs themselves?

Photocopying.   Will you have any copying costs? Certainly you will want to make copies of the final format, script or storyboard and certain of your planning lists for your cast and crew.

Set.   Will you need to construct any portion of the set? Large or small? Or will you need to rent some part of the set?

Cast.   Will your performers have expenses that need to be covered? Transportation, food, or overnight accommodations, for example. In professional TV productions, the principal cast members and sometimes some of the crew members receive a per diem on the days they work. A per diem is a daily fee paid, say ten dollars or fifty dollars, to cover all expenses including transportation and food but not lodging for that day. Lodging is covered separately. With a per diem, neither the producer nor the cast or crew member has to worry about receipts. It not only simplifies covering cast or crew expenses but it puts a limit on these expenses. If they spend more than the designated per diem, that comes out of their pocket, not yours. You may or may not want to do this.

Dub Cost.   When you dub the final show, what will it cost per tape? Multiply this by the number you expect to have to make to determine an estimated cost. If you are doing the dubbing on your own equipment, you may elect to charge only the cost of the tape itself. You could even ask all those who want a dub to supply their own tape. In this case, you would not need to budget a cost for dubbing. In either case, however, remember your time is worth something and don't be bashful about charging for it. An additional five dollars per tape, for example, is not unreasonable. Whatever, you should make a decision about how you are going to handle dubbing and put this in the budget.

Wardrobe.   Will any special wardrobe be required by the cast? If you are responsible for getting this wardrobe, it should be in your budget. In our case, we are not supplying the wedding gown so we will not include this in our budget. If, however, we were asked to provide a special kind of bow tie for a sequence with the groom, we would include the cost of this if we were required to purchase or rent it. Your wardrobe requirements may be limited to special items only with the bulk of the videos using the everyday street wardrobe worn by your cast.

Makeup/Hair.   Will any special makeup be needed? A mask? Body paint? Fake mustaches? Will any special hairdos be needed? Wigs? Makeup and hair are sometimes grouped together like this with the same person responsible for both. Will you need to budget for these two items?

Miscellaneous Expenses.   This is your safety column, for all those expenses you may have forgotten or those costs you may have underestimated. We suggest that you total your budget and add 10 percent of that to the final total just to be safe. A 10 percent cost overrun is not without precedent in the business world. This safety measure may save you explanations and embarrassment later.

Your Fee.   Are you going to charge for your time and expertise? Certainly if you are doing the video for someone else, you should consider this, even if that someone else is a friend. For example, we may elect to charge for our time on the wedding video or we may make our time in doing it our wedding gift to the bride and groom. If you do elect to charge for your time, calculate this either by the hour or at a flat fee for the video. We encourage the latter, a flat fee, since it easier on everyone. You do not have to keep track of your time and the person paying for the video knows exactly what you are costing. There are no surprises when the video is delivered and final payment is made. In estimating your worth, use your present salary as a guide. What is your time worth in the real world? Try to estimate your time input as accurately as possible so the flat fee you calculate is representative of your time commitment.

Again, a budget is not a final cost. It is as it is titled, just a budget. It is used only to calculate approximately what the video will cost and, if you have an executive producer, it is a means to justify an advance of funds to begin the process. It is common for the executive producer to advance 50 percent of the BUDGET to begin the production planning, with another 25 percent being paid on the first taping day and the final 25 percent paid on delivery of the video. You may want to adapt this schedule to suit your own needs and financial situation as well as the circumstances under which the video is being produced. Whatever you do, make sure that the executive producer is covering the expenses as they are spent. You should not have to finance someone else's show.

Once the video is complete, you will take your original budget, write in the exact costs opposite the estimates and calculate the final cost less any advance. Present this along with your receipts to the executive producer for final payment.

After you have prepared a budget several times, you will begin to get very good at determining offhand how much it will cost to produce a show. If you are the executive producer, you may discover that those videos you were playing around with are actually costing you quite a bit of money. All the more reason to make them good ones.

The budget for the wedding video is shown in Figure 4–2 for your further understanding of the budgeting process.

CONCLUSION

Making a budget for a video will likely open your eyes to hidden costs that you may not have realized you were incurring. The excitement of creating a video sometimes overwhelms the production and the actual cost of doing it gets lost somewhere along the way. A budget is a good way to put the project into perspective before you run off to do it. Once the budget is complete, the obvious question is: Is this video worth that amount of money? If it is, then move on to the next step. If not, rethink your idea. Maybe there is a better one that makes more sense financially.

Coming up next: *The Equipment and the Crew* . . . who is going to do what with what.

TO DO

1. Make a budget.
2. Check the budget off your Master Checklist.

Show Title Wedding Video

Date _____

BUDGET WORKSHEET

Note: Estimate all costs. It is not necessary to make any final decisions about
locations, equipment, crew, edit, or set at this time. This budget is a best-guess
estimate of all expenses for the proposed video.

VIDEOTAPE

    Cost per tape $ 6.00

    No. Cassettes   ×   11

                  Total $ 66.00

LOCATION FEE

    (Place, if you know)

    _____ $ 0

    _____

    _____

                  Total $ 0

TRANSPORTATION

    Gas $ 20.00

    Parking   5.00

    Toll charges

    Airline tickets

    Misc.   15.00

                  Total $ 40.00

PROPS

    Buy $ 0

    Rent $ 0

                  Total $ 0

Figure 4–2   Wedding budget.

EQUIPMENT

(Name it, if possible)

Video _____

_____

_____ $ _____0_____

Audio _____
microphone

_____   20.00

Lighting _____
camera mounted

_____   20.00

Misc. _____

_____   Total $ 40.00

CREW

Cost per $ 20.00

No. crew members   × _____8_____

Subtotal   _____

Plus other   + _____

Total $ 160.00

EDIT

Rent/buy equip _____

_____ $ _____

Subtotal $ _____0_____

Postproduction facility

cost per hour $ 10.00

No. hours est.   × _____2_____

Subtotal $ 20.00

Total $ 20.00

Figure 4–2 (*cont.*)

TELEPHONE

    Estimated cost $ __5.00__ $ __5.00__

ARTWORK/GRAPHICS

    Misc. expenses $__20.00__

    Artist   _____

         Total $ __20.00__

CATERING

    Estimated cost $ _____ $ __Ø__

MUSIC

    Buy records/tapes $_____

    Original music $_____

         Total $ __Ø__

ADDITIONAL VISUALS

    Photos $_____

    Video   __10.00__

    Film   _____

    Other   _____

         Total $__10.00__

POSTAGE/DELIVERY SERVICE

    Stamps $ __Ø__

    Delivery $_____

         Total $ __Ø__

PHOTOCOPYING

    Estimated cost $__10.00__

    Supplies   __5.00__

         Total $ __15.00__

SET

    Planning $ __Ø__

    Construction   _____

         Total $ __Ø__

Figure 4–2 *(cont.)*

CAST

    Expenses $ _____ ⊖ _____

    Per diem

                  Total $ _____ ⊖ _____

DUB COST

    Cost per tape $ 6.00

    No. cassettes   ×   10

    Subtotal $ 60.00

    Dubbing fee $ 5.00   } gift

    No. cassettes   ×   10

    Subtotal $ 50.00

                  Total $ 60.00

WARDROBE

    Buy $ _____ ⊖ _____

    Rent

                  Total $ _____ ⊖ _____

MAKEUP/HAIR

    Supplies $ _____ ⊖ _____

    Misc. expense $ _____

                  Total $ _____ ⊖ _____

                             $ 436.00

TOTAL ALL

MISC. EXPENSES

    Total all $436.00

    Add 10%   ×   .10

                  Total $ 43.60

YOUR FEE   gift $ 100.00

TOTAL BUDGET $479.60

Figure 4–2 (*cont.*)

CHAPTER FIVE

THE EQUIPMENT AND THE CREW

Materials needed for this chapter:

Completed

SHOW OUTLINE

Still Working On

MASTER CHECKLIST

Blank Forms

EQUIPMENT LIST
CREW LIST

Now that we have in hand the beginnings of the video, we can start to nail down certain specifics. There are two ways to go here. We can begin with our equipment and crew lists or we can go from the outline directly to the format and script. Some would say that you cannot begin a formal format or script if you have not determined what equipment and crew is available. Others would say just the opposite, that you cannot possibly decide what equipment and crew you will need if you do not have a format or script. Both arguments are valid. However, our experience has been that it is best to define the parameters of what is immediately available in terms of the equipment and crew and then to proceed with the format and the script. It is far easier to write the format and script to fit the available elements than it is to redesign or change the video after you have written it beyond the limits of the equipment and crew. We have elected, therefore, to do the equipment and crew lists next and then follow with the format. As you work on various videos, you may discover that you prefer the opposite approach. That will be a part of developing your own style.

Throughout this chapter, you will want to consult your show outline completed in the previous chapter as you begin to lock in specific plans for your video.

WHAT EQUIPMENT IS AVAILABLE TO YOU?

First, what equipment do you have? What can you borrow or rent or promote or whatever for this particular video? Will you need it all every time you tape? You will most likely have one day where all equipment must be available to you. Ideally, what would you like to have on that day? Let's break it down into specific types of equipment and write it on an equipment list as shown in Figure 5–1.

The Cameras

Obviously, you will need one or more cameras. Two is better than one; three is better than two; and four is even better than three or so you would think. This, however, is not always true. While you do want to get the shots, you do not want to overload yourself with too many cameras. Remember that every one of your cameras will create a piece of tape that must be integrated with tape from other cameras. This could mean long hours of previewing tape, choosing shots, preparing the edit list, and then spending time in the edit inserting and removing tapes from the VCR, cueing up tapes, and making shots match.

By making shots match we mean matching the colors being recorded by each camera. One camera, for example, may see one shade of red while another camera sees a different shade. Getting perfect matches from one camera to another may be difficult, particularly if the cameras are different brands or the same brand but different models. All cameras will take a slightly different picture. Some of this depends on how a camera is set up before it rolls tape. A subtle color difference between cameras, for example, could be the result of how they were white-balanced. Did you use the same source of white under the same lighting conditions? When you try to cut video together from two cameras that don't match, the viewer will be able to see the change in the picture quality. This could, of course, affect that viewer's opinion of your abilities as a producer.

The bottom line here is that multiple camera shoots are preferable to single camera shoots but they have their own special problems to which you must be willing to adjust.

All that aside for the moment, how many cameras do you think can capture what you have in mind? Will one do? If so, then your camera problems are solved. You have one, right? Right. If one will not do, then do you own more than one or can you borrow or rent others? Maybe you have a partner in this venture and that partner owns a camera or knows where one can be borrowed or rented.

Show Title _____

Date _____

EQUIPMENT LIST

CAMERAS

    Total no. _____

    Format (specify number)

    VHS _____   Beta _____   8 MM _____

    Designated duties

        #1 _____

        #2 _____

        #3 _____

        #4 _____

MICROPHONES

    Total no. _____

    Type (Specify number of each)

    Handheld _____   Bo-om _____

    Mike stand _____   Lavaliere _____

LIGHTING

    Total no. units _____

    Type (Specify number)

    Pro lamps _____   Shiny board _____

    Camera mounted _____   Other _____

TAPE

    Total no. cassettes _____

    Kind (Specify number)

    VHS _____   Beta _____   8 MM _____

Figure 5–1 Equipment List form.

TVs/MONITORS

(Number and size of screen)

No. Black & White _____

No. Color _____

ACCESSORIES

(Specify number)

Tripods/stands

Camera _____   Lighting _____

Audio _____   Dollys _____

Lighting gels (color and number of each)

color _____   # _____

color _____   # _____

VCRs

Total no. _____

Kind (specify number)

VHS _____   Beta _____   8 MM _____

Figure 5–1 (*cont.*)

When you are trying to decide how many cameras you ideally want, ask yourself one major question, "Do you have control of what you plan to tape?" In other words, can you get your performers to redo a scene if you do not get a good shot? Or, as in the main part of our wedding, do you have one chance only to get the shot and no opportunity to reshoot? The answer to this question should be considered when you are deciding how many cameras you want to use. It would be wise, if you do not have control, to roll at least one backup camera, perhaps locked on a wide shot. It will always be recording a shot you can cut to or use entirely if the first camera does not get the action for whatever reason. You could, for example, be shooting tight on the bride and groom as they exchange their vows and someone could move in front of your lens blocking out the scene entirely. If you have no backup camera, you have nothing to insert over this mistake. Not to mention you will be losing a major part of the event.

Also, you will want to take into consideration the way you see the video. Do you see wide shots of the family and friends gathered for the wedding? Do you see tight shots of the mother of the bride brushing away a tear or the flower girl throwing rose petals in the path of the bride? To get these shots just mentioned, you would need more than one camera. Think this through.

For our wedding video, first we need a camera for the primary action and the major performers, the bride and groom. Because of the importance of the event, ideally this is a camera solely devoted to them so that we will not miss a moment of their reactions throughout the event. We will call this primary camera Camera #1.

Next, we want a camera that can spend all or most of its time on a wide shot of the location, whether church or reception hall. This wide camera will supply perspective for the closeups taken by Camera #1 and it will serve as a backup as well. This wide camera does not necessarily need to be manned. It could be mounted on a tripod, positioned to capture a certain area, then locked off and left. While this is not preferable since it gives no flexibility to the shot the camera is recording, it is a way to record the shot without an additional crew member. Even if you decide to do this, you will need someone to start the tape rolling and to insert new tape as it is needed. If you have a two-hour tape, this is not a time consuming job. We will call this second camera our wide camera, designated Camera #2.

Finally, we would like to have a camera devoted strictly to reaction shots. This camera would require a crew member. It would float, moving to capture tears, laughter, congratulations, and anything else going on at the event. The video, after all, is not just about the principal performers. It is about everyone there down to the most minor role. It is this combination of all the experiences and reactions that will make the video successful. We will call this our floating camera, designated Camera #3.

We have shown where we might position these cameras in Figure 5–2. You do not need to draw a sketch like this now. We will cover this later. We show it only to

Figure 5–2　How many cameras will you need and where will you place them? In this example, Camera #1 is handheld; #2 is mounted on a tripod; and #3 is handheld.

demonstrate how you need to think through the question of how many cameras are enough by positioning them in your mind at their designated locations.

You could add another wide camera to cover a different angle or you could add another floating camera to cover more reaction shots. There comes a point, though, where there are too many cameras and all you will end up with are cameras shooting cameras. That you certainly do not want. You never want to see one of your cameras in the video. That is the sure sign of an amateur unless, of course, it adds to the story line and is a planned part of that video.

As you can see, in planning cameras for our wedding, we have not just arbitrarily said, "I want three cameras." Rather, we have chosen cameras based on the way we see the show, assigning every camera a specific function.

Camera #1.   Primary camera
    Duties: major performers, closeups
Camera #2.   Wide camera
    Duties: wide shot
Camera #3.   Floating camera
    Duties: reaction and crowd shots

We have also numbered every camera. This number will be used to call for certain shots and all tape and log sheets from the cameras will be labeled, as in Figure 5–3, with the title of the show, the camera number, the cassette number for that camera, the day, and date. This information will make it easier to locate material later. Things can get confusing fast, particularly if you have multiple cameras. Say that there are three cameras using three cassettes each for a total of nine cassettes for the shoot. With no tape identification and no logs, finding a shot could be so time consuming as to make it impossible. The bottom line: *label the cassettes*.

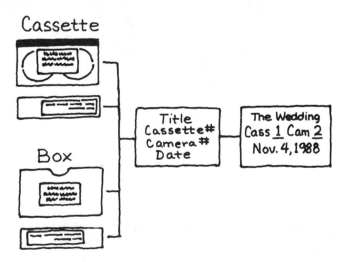

Figure 5–3 How to label the cassettes as you use them.

When we get down to planning the edit, the ideal situation would allow us to have a VCR and TV for every camera and to roll them simultaneously so that we could see which camera had captured the best shot at that particular time. This, of course, assumes that all cameras are rolling all the time and this is not always true, so this will not always work. In any event, we will get to that later when we discuss editing. We mention it now only to get you to thinking about the relationship between what you actually record on tape and how you make it into a video that works.

If you were shooting our wedding, you might want to add cameras beyond the three we have decided we need. You may decide you don't even need this many. You may not have access to this many. The idea is to define what you would like and then to make that match to what you really have available.

Write in the number of cameras and their designated duty on the equipment list.

Camera: #   3

    Specify number: VHS   3   BETA _____   8MM _____

    Duties: #1 Primary

          #2 Wide

          #3 Floating

If you cannot get as many cameras as you like, begin cutting with the least important camera. In the case of our wedding, this is Camera #3. We can, after all, live without a special reaction shot camera. We could tape the show with only two cameras by allowing the wide camera to tape wide part of the time and to tape reaction shots part of the time. This simple adjustment could eliminate a camera, a possible crew member, and extra tape that must be viewed and edited. Hopefully, you will be able to have as many cameras as you want but, if you cannot, then adjust your thinking so that you still capture all the action.

One last thing. It is not important that you have all the same make or format of camera in a multiple camera shoot. You could, for example, have an RCA VHS camcorder and a SONY BETA camcorder on the same shoot. It will not matter because each will have its own recorder built into the camera. It will matter only in the edit and even then it will not be a huge problem. It means only that you will have to tie together a VHS and a BETA for the edit, which is generally no big deal.

You should, however, be sure to check to make sure your editing system can handle the different kinds of cameras and tape. Also, be sure to check out different cameras prior to the shoot to make certain that the pictures match; that is, that the quality of picture being recorded by one camera is equal to the quality of picture being recorded by another. If they do not match, then as we have already mentioned, when you cut from one to another, the difference will be obvious and looked upon by the audience as a mistake on your part. In TV, the buck stops with **you,** the producer!

The Audio

There is a tendency to take audio for granted. When we think of television, we tend to think of it as a whole, not as a combination of both video and audio. If you are not careful, this same lack of recognition of two of the major elements that make videos will show up in your work as major audio errors. It is true that some cameras have built-in microphones and, therefore, when you roll tape you record audio also unless you specifically turn the audio off. But it does not follow that you can forget about audio entirely and trust your camera to deliver a perfect audio track. Good audio, like the camera itself, takes careful planning.

First, take into consideration where the camera is located in terms of the sound you wish to record. It would not be unusual that the sound would be too far away to record. You can, after all, tape your picture from a distance and record both wide shots and closeups using the zoom on the camera lens. However, you cannot adjust your built-in microphone to the distance. It will record only within a certain area, as shown in Figure 5–4, and that is it. No more and no less. You cannot zoom in your microphone. You may discover, therefore, that the camera's microphone may not be enough to record all the sound you hear for the video. You must plan your audio requirements just like you plan your cameras or you may end up with great pictures and lousy sound.

The impact of this can best be demonstrated by turning on your television with the sound off. Without the audio, the pictures lose their specific meaning and become a puzzle that you must decipher. The puzzle becomes the focus of your attention and the show is lost in the process. Bad or patchy audio is even more distracting and irritating to the viewer.

As an example, let's take a look at our wedding. Imagine our wedding taped with only the built-in microphone. If our primary camera is close to the altar during

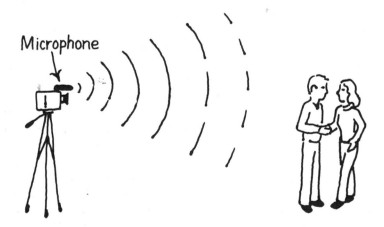

Figure 5–4   Recording audio with the camera microphone. The camera's built-in microphone will record audio only within a certain area.

the ceremony, we will have acceptable audio but that audio level will change every time we move. Say we follow the bride down to the altar, and then settle in for closeups. As the ceremony proceeds, we may move this primary camera behind the preacher to better record the action. As we move, we will vary our distance from the principals and the audio will fluctuate up and down. You could even lose audio entirely while you move from one side to the other.

Think of it like this. You are at a concert and want to take a photograph of the performer onstage. It is dark inside, so you use your flash. Wrong. The flash of light generated by the flashbulb will not travel fifty feet to the stage and it will not help to light the subject. You may see a great picture in your viewfinder but you will not be able to take it with your flash. The built-in microphone works the same way. It will only record sound up to a certain distance. After that, it needs some help or what you hear will never get to the tape.

Every camera on the shoot must be designated an audio feed. In some cases, the built-in microphone may be enough. While our primary camera may need more than the built-in, our wide camera with its built-in microphone can record ambient sound very well. That is, it is recording all the sounds surrounding the ceremony including crowd and music. Our floating camera, on the other hand, may need help depending on what it is recording. Reaction shots will need no more than the camera mike while interviews may require a handheld mike for the best sound quality.

The kind of sound you want to record is different throughout the video. Sometimes ambient sound is enough. Sometimes it is not. Again, it is important to think of your audio as separate and aside from the picture when you are planning.

Think of your audio options as shown in Figure 5–5:

1. BUILT IN MICROPHONE.   The camera microphone.
2. BOOM MICROPHONE.   Plugged into the camera and mounted on a long pole that is held by a grip over or below the performers out of camera frame.
3. HANDHELD MICROPHONE.   Held by the performer and plugged into the camera.
4. MIKESTAND MICROPHONE.   Positioned in front of the performer and seen in the shot. Plugged into the camera and repositioned if the camera moves.
5. LAVALIERE MICROPHONE.   Worn by performers. Wired to the camera.

For our wedding video, the answer to our audio dilemma is simple. Because we want to capture every moment and that includes the sound, we will need at least one microphone not attached to the camera, a remote microphone. We may decide to add even another later. For now, we would position this one at the altar where the ceremony will take place, probably on a mikestand so it will remain stationary. This remote microphone must be cabled to the camera. It will, therefore, be necessary to keep the camera within cord length or to plug and unplug the mike as it is needed. A plug–unplug situation is undesirable but perhaps necessary. If, for example, you

Figure 5-5 Kinds of microphones.

want to follow the bride down the aisle with the primary camera, the remote microphone is not necessary. If, however, you want to then focus on closeups of the principals during the ceremony, the remote microphone is necessary to get good clean audio. After the ceremony is over, we will convert this stationary microphone into a handheld microphone by taking it off the mike stand so we can record special interviews or happenings during the reception.

As far as the other two cameras on our shoot go, their built-in microphones will do fine for the wide shots and the reaction shots.

> *Microphones #* __1__
>
> Specify number: Handheld _____   Boom _____
>
> Mikestand __X__   Lavaliere _____

We do not want to confuse the issue, but to keep the thinking process going, you should remember that every camera that tapes audio will give you additional audio options when you edit. You may find it reassuring to know that you do have additional audio available on another piece of tape should you need it. Audio can always be separated from the picture it was recorded with and used behind a picture recorded on a different camera. Just as you can use different pictures from different cameras when you are editing, you can also use different sound from different cameras.

You may decide that you do not need to tape audio on every camera. You may feel like you have enough. The tape from a camera not recording audio should be labeled MOS, meaning pictures with no sound. An MOS shot of a sunset, for example, might be added to help create the mood of the close of the wedding video. You don't need the sound to that sunset but rather you will probably be adding music to enhance the effect. We want to make it clear though that we do strongly recommend that you record sound at all times. It is always better to have it and not need it than the reverse.

We have planned our wedding video so that the main audio track will be on one camera; in this case, the primary camera #1. Camera #2 and #3 audio will be used as backup. By putting our main audio track on one camera, we will make the edit easier and faster, eliminating a lot of cassette changing and searching, not to mention confusion over which audio to use.

Now on your equipment list, designate your audio requirements. You need not list the built-in camera microphones but only those extra mikes you will need.

The Lighting

How do you plan to light the video? Perhaps you believe that your camera needs no additional light. This may be true for certain situations but totally wrong for others. The best idea is to test your camera in the exact location to determine for sure what it can see and record. Generally, a camera can record about one-half of the light you can see with the naked eye. It may not, however, record it as well as you can see it. Your eye adjusts for different levels of light in a particular scene. The television camera flattens the scene and records the levels as one with the darker areas falling off and tending to be blacker than what you see. So what you believe is going to be a great picture may be gray on the tape.

Television cameras need a lot of light to "see." That is why there are many shows on broadcast TV that are flooded with light. No shadows are allowed because, as we pointed out, these become blacker in the camera. On the other side, you will see shows on TV that *use* shadows to add to the storyline, like *Miami Vice*. Be careful not to use what you see on television as your guide to lighting, particularly shows like *Miami Vice*. It is shot on film, not video, and the lighting requirements of a film camera are totally different from what they are for a video camera. Just because a show is on TV does not mean it was shot on videotape.

You may or may not need additional lighting for your video. The ultimate determining factors will be the capabilities of your cameras and the locations where you will be taping. The easiest way to answer the lighting question is to take your camera to the exact locations *now* during the planning stages. Roll tape with the light as it will actually be on the day or at least as close as you can get it. Then review this tape to determine additional lighting requirements. If you require additional light, you can rent or borrow professional equipment or you can make do with everyday lighting instruments available at the location. Something simple may be the answer. Use your imagination. Consider these ideas for answering your lighting needs:

1.   Sometimes a lamp with the shade off will add just enough extra light. If you put a brighter bulb in that same lamp, you will have even more light. And if that lamp can be turned in different directions or is mounted on a stand, it is even easier to direct the light.

2.   Any photographic lighting equipment may work except strobes, since their light is too short in duration to make a lasting impact on a scene.

3.   A high-intensity flashlight will even add light to the scene—not a lot but some, and perhaps effective light at that. There are all kinds of flashlights and many

different intensities of light. Light from a flashlight is mostly focused in one area and this may create an unwanted hot spot. Watch out for this.

4.   Reflected light may be the answer. You can use sunlight or a light source across the room by making a shiny board, a piece of cardboard covered with a shiny material, say tin foil. You use this board to bounce light off the reflective surface onto the subject. For example, sunlight reflected off a shiny board will spot the subject it is directed onto, as shown in Figure 5–6.

Figure 5–6   The shiny board directs light to the subject by reflection off its shiny surface.

5.   You can buy or rent special video lighting equipment. An outright purchase might make sense if you plan extensive use of your video equipment. A camera light, one that is mounted on the camera, would be a good place to start. Otherwise, rent or borrow what you need.

6.   Remember to use the lighting adjustments on your camera; one is shown in Figure 5–7. Every camera has the ability to adjust to the situation and perhaps that alone will solve most of your lighting problems.

It doesn't hurt to use your imagination. Actually, that is probably the best idea since it will force you to test the lighting situation and begin to discover how different lighting and the lack of sufficient lighting will affect your video productions. You cannot make any assumptions about light. You cannot, for example, assume that since you are taping outside on a sunny day that your lighting will take care of itself. You could have all the light in the world and your picture could be lousy. Taping a person in strong light creates shadows on the face, particularly under the eyes. You

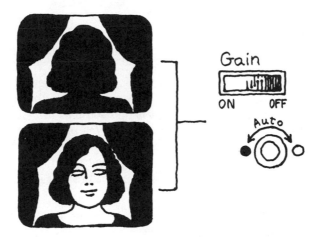

Figure 5–7 Your camera will have some ability to adjust to the lighting conditions at the recording site.

could end up with the most gorgeous day ever and your performer standing there with what looks like two black eyes.

Without trying to give you all the whys and hows of lighting here (that is a whole book by itself), we can only emphasize the importance of lighting and being aware of what kinds of light are available at your locations. Again, we recommend a test tape at the location and experimentation with light.

For our wedding video, we have completed our test and discovered that we will need some additional light at the altar. As you remember, we particularly want to capture this action. We have decided to add two gooseneck lamps with high-watt bulbs positioned to either side. We have tentatively decided where they will be positioned based on our test, but will fine-tune this on the day of the taping. In addition, we will need a light on our floating camera. To make this simple, we have decided to mount one on the camera itself. Specifically, we will be using a light designed for this purpose. In this way, at least in the closeups, we will be assured of additional lighting. For our wide shots, we will rely on available light. We would add these lighting elements to our equipment list.

Lighting # _____3_____

Specify number: Pro Lamps ___2___   Shiny Board _____

Camera ___1___   Other (name) _____

The Tape

You will need tape. This is an element taken for granted. It should not be. You should give this more than just passing thought since your video is nothing without it. First of all, you should consider the quality of the tape. Assuming that you are taping a video that you want to last beyond next Thursday, you want the best tape you can buy. Spend the extra bucks because it could mean better pictures, better sound, and the high-grade videotape will stand up better against repeated plays while you are recording and editing. You particularly want a top-quality edit master to make your dubs from. If you use a poor-quality tape with dropout problems, those problems become a permanent part of your video. If the material is not recorded properly, it simply is not there. So it makes sense to record and edit on the best. If you want to save money on tape, do it when you make your dubs.

The other question to be answered is: How much tape will you need? You will have already estimated this on your budget, but you do not want to run short when you need it most, so consider the following before you run out to buy tape:

1.   RECORD SPEED.   Camcorders record at only one speed so if you are using only these, forget this question. If, though, you are using a separate record machine, a VCR for example, you may have the option of recording at different speeds. If you do, select the standard setting. It records a higher quality picture.

2.   TAPE LENGTH.   You will want the shortest length of tape you can buy. The whys of this are simple. If you have two hours of original material on one cassette, you will conceivably have to wait an interminable length of time to search from front to back. This will add to your review time and your edit time, not to mention the fact that it will be excruciatingly boring. It is best to have short pieces of tape and more cassettes than the opposite. While this may cost more money initially to purchase the additional cassettes, it will save money in the edit. And, remember these original material cassettes can be erased and reused for something else once the video is complete and edit master created.

3.   HOW MUCH.   Decide approximately how much tape you believe you will need. Then plan to buy more than that amount. You do not want to run out and you can always use it on another shoot or you can return it to the store for a refund. It has been said that, "It is better to be safe than sorry." That goes double on a video shoot.

For our wedding video, we have decided that we will need two cassettes per camera so we will buy three per camera to be safe. So, we would write on our equipment list:

TAPE #___9___

VHS ___9___   BETA _____   8MM _____

It is always a good idea to buy the same brand of videotape for the shoot and the edit. Quality does vary from one manufacturer to another and this is just a good practice for further insuring the quality of your video. Assess your tape needs, add to them for safety's sake, and write this number on your equipment list now.

TVs/Monitors

Will you want any TVs or monitors at the location during taping? This is a good idea if you can find a place to set up one or two. A camcorder will play back only in black and white in its viewfinder and while you can check framing and picture composition in black and white, it is not possible to get a true feel for the picture being recorded without seeing it in color.

It is desirable then to set up at least one TV or monitor at the taping location in a quiet place away from the activity. You can then go there to view tapes, as time is available or as you feel the need. Another option is to set up this TV or monitor right next to the camera so it can be seen at all times. A potential problem with the latter is that when the camera moves, the TV or Monitor must move also. This will not be a problem if you have the camera mounted on a dolly. However, it could be a problem for a floating or handheld camera. Where would you put the TV or monitor then?

Also, having the TV or monitor there all the time may get the camera operator into the habit of using the TV to see the shot rather than the camera's viewfinder. This is not a good habit to acquire. It will limit the operator's movement considerably having to drag along a TV all the time. It is better to use the camera the way it was designed and plan a quiet place to set up your TV or monitor away from the activity, in a viewing room where you can view and evaluate the recorded material on an ongoing basis. More about this later.

There is also a variation on the TV/monitor-at-the-camera-site that is not only acceptable but may be just the thing for you to continually monitor your camera operators. It will require some work on your part though. Whenever you want to see a camera shot at the camera location, you will need to carry a small TV/monitor to that location, plug in and play back. Preferably this is the smallest color TV/monitor you can get, 5 inches or even smaller, and preferably the unit is battery powered so you do not have to plug in to an electrical circuit. This portable setup will allow you to see the recorded material at any time as well as check out the camera framing and moves in color. Try not to let this replace the proposed viewing room; let it supplement it. Such a portable monitoring system could make you more effective as the producer/director, giving you more control over what is recorded.

For our wedding video, we have chosen to have both a viewing room and the portable viewing setup. So, on our equipment list for our TV/monitor requirements, we have written:

TVs/Monitors (number and size of screen)

Black and White _____

Color <u>1 19″ and 1 5″</u>

VCRs

If you set up the TV or monitor as suggested in a quiet place, it would be nice to have a VCR dedicated to this viewing room so you do not have to pull a camera off the shoot to use it as a playback machine. Also, a VCR dedicated to this setup will be plugged in and ready to play immediately at all times. Your camera can stay in position and even record new footage while you view what is already on tape.

We do plan a separate viewing room for the wedding video, as you will remember, and we do want a separate VCR devoted to that room so to our equipment list we would add:

VCRs

Total #   1

Kind (specify number)
VHS   1   Beta   8MM

Since all our cameras will be VHS format, we will want a VCR that is VHS also.

Accessories

What other equipment will you need to support the equipment you plan to use? Tripods? Light stands? Mike stands? A dolly for one of the cameras? Colored gels for the lights? Special plugs or cable or whatever? Make a list of what you believe you will need.

For the wedding video, we have already decided that Camera #1 and Camera #2 will need tripods and that we will want one microphone mounted on a stand. We do not plan to mount a camera on a dolly at this point. So on the equipment list, we would add:

Accessories (specify number)

Tripods/stands
      Camera   2   Lighting   Audio   1

We have not given a great deal of thought yet to the effect a gel might have on a scene. Perhaps we might want to try some to see what we can do with them to add to the show. Since this is a happy but emotional event, let's choose a blue. We will try it out over the lights at the altar to see if we can create a more romantic look. We would write on our equipment list:

Lighting gels (color and number of each)
      color   blue   #   2

Gels can add to or distract from a scene. If you are considering the use of one, try it out beforehand to make sure it is creating the effect you envisioned.

Once your equipment list is complete, you can begin gathering all the items, adding or deleting as needed. By the first taping day, the list should include absolutely everything you will be using and it can, therefore, then become a checklist to make certain all the equipment is at the location on the given day.

WHAT CREW IS AVAILABLE TO YOU?

Behind every great performer there is a great director, a great camera operator, and maybe even a great grip. You could have a superstar as your principal performer and if the crew is lousy, the superstar will probably come off lousy, too. Television is teamwork and everyone on the shoot is important to the video's success. A sloppy crew will make for a sloppy video. What your crew lacks in skill, however, it can make up in enthusiasm. If the crew cares about how the video looks, they will take more care in not only their duties but in the duties of others.

This will be particularly important to you since you will not have absolute control of your cameras once they are rolling tape. If you are not manning a camera yourself, you will be able to check in on everyone to make sure they are following the plan you have developed; you may even have a chance to play back bits and pieces of what has been recorded on a Portable Viewing Setup or in a Viewing Room. The ultimate control, though, will remain with the camera operators. Only they will know whether what they are recording is framed right, lit right, and works in terms of the total video production. You should, therefore, cast your crew just as carefully as you cast your on-camera performers.

The Possible Crew Positions

Using the form shown in Figure 5–8, let's assess our crew needs.

The Director. The director plans the production according to the concept, format, and script. These plans include breaking the script down into specific shots, locations, and effects; hiring the crew; designing the style; directing the performers and the crew; and finally supervising the edit to create the video as it was conceived.

The Lighting Director. The lighting director (LD) plans the lighting and then makes sure that the instruments are set up properly and that the lighting works when seen on tape.

The Technical Director. The technical director (TD) makes sure that the picture is technically good. This crew member also pushes the buttons according to the director's call during a professional switched camera shoot. For the simpler

Show Title _____

Date _____

CREW LIST

DIRECTOR _____

LIGHTING DIRECTOR _____

TECHNICAL DIRECTOR _____

AUDIO DIRECTOR _____

SET DESIGNER _____
SET CONSTRUCTION _____

Figure 5–8 Crew List form.

VIDEOTAPE OPERATOR _____

CAMERA OPERATOR #1 _____
CAMERA OPERATOR #2 _____
CAMERA OPERATOR #3 _____
CAMERA OPERATOR #4 _____

GAFFER _____

GRIP _____

MAKEUP ARTIST _____
HAIR STYLIST _____

Figure 5–8 (*cont.*)

WARDROBE _____

PRODUCTION ASSISTANT _____

_____

ASSISTANT TO:

CAMERA 1 _____

CAMERA 2 _____

CAMERA 3 _____

CAMERA 4 _____

GO-FER _____

Figure 5–8 (*cont.*)

shoot, this crew member can also man the viewing room, keeping it up and operating, ready for use at all times.

The Audio Director.   The audio director (AD) handles the design, placement, and operation of all microphones. The AD may also suggest the use of any sound effects or music.

The Set Designer.   The set designer designs how the location will look at the time of the taping and creates the plans for the construction of a set for the taping.

Set Construction Crew.   This crew builds any special set pieces to the specifications of the set designer.

The Videotape Operator.   A videotape operator controls the record machines that are remote from the camera, like a VCR. This crew member also continually checks picture and sound quality to make certain both are recording properly.

The Camera Operator.   Number one on the list when casting this crew member is experience. Enlist someone who has planned and made their own videos. If possible, have them use their own camera. You will then have a trained camera operator who knows the equipment. If you cannot get someone with experience, choose someone who is anxious to learn not only how to run a camera but how to make a video. This person will give more attention to detail and, hopefully, not be afraid to ask questions when she or he does not know what to do. While the temptation may be great, try to resist the urge to make yourself a camera operator. Things will go more smoothly if you can function as the director, moving from one camera to the other to check shots and preserve the integrity of the concept of the video. It is, after all, your video and no one—no matter what you do—can see it the way you do.

Gaffer. A gaffer positions lights according to the lighting director's plan and then makes certain that these lights run properly during the taping. If your additional lighting needs are minimal, you may not want to tie up a crew member just on this but may elect to combine this duty with others.

The Grip. A grip lays or pulls camera cable and helps in all aspects of the setup. If you have a handheld camera with a remote microphone, you may want a grip to carry the microphone, making sure that the cable does not get tangled or hung up as your camera operator moves about. You may need more than one grip.

Makeup Artist. The makeup artist designs and applies makeup.

The Hair Stylist. The hair stylist may design hair styles appropriate to the roles. This crew member also keeps the hair looking the way it should during taping.

Wardrobe. This crew member gathers and in some cases chooses the clothing to be worn by the performers. Wardrobe also makes sure that all clothes are clean and pressed and at the taping site ready for use.

Production Assistant. Sometimes called *assistant to the director,* this crew member is vital to your shoot. His or her number-one duty is to take notes on the production. These notes will be crucial to you when planning the edit and during the editing session. Choose someone who not only can take copious notes but someone who knows you well enough to fill-in-the-blanks when you are too rushed to complete a specific instruction.

Assistant to the Camera Operator. This is not a usual position on a professional shoot. However, since you will not be able to monitor your cameras all the time, you must have an assistant for every camera to take notes. Otherwise, you will have to view every single piece of tape from every single camera repeatedly to find shots. A camera assistant can take notes based on the camera operator's observations of the video being shot, noting the better shots, the ones not to be left out, and those that just don't work, and noting where all shots are located on the tape using the camera counter number. This person can also make sure that the cassettes are labeled properly and can generally assist the camera operator in moving from one place to another.

The Go-fer. This is generally considered the lowest position on the totem pole in a television shoot. The name go-fer comes from the duties of this person who "goes for" anything anyone asks for. This crew member does whatever is asked. Generally, this person is kept extremely busy on the taping day running from here to there doing things that are totally unplanned like going to buy more tape or a replacement for a light bulb that has burnt out. This person should of necessity be energetic and preferably young.

Choosing a Crew

The size of your crew will be determined by the size of the taping. More cameras usually mean a bigger crew. A more elaborate set, audio design, or lighting design requires additional crew who can devote their time to their own specific area so that the director can be free to oversee the total production. If you do not designate someone to do certain crew duties, that does not mean these duties will not be done. On the contrary, it means that *you* as the director will do them or that you will delegate them to another crew member. Just because you don't have a lighting director does not mean no one has to light the scene. It still has to be done.

Choose your crew with care and above all, treat them like professionals from day one. If you *demand* the best from them, they will be more likely to deliver it. If you treat them like amateurs, they will respond in kind. Demand that they attend preproduction meetings, that they report for rehearsals on time, and that, on the appointed days, they treat your video just like they would if they were working on an episode of a prime time network TV show. Demand the best and accept no less.

On your crew list, write in those crew members you will be needing and the names of likely candidates. For our wedding video, we have written in:

Director	You
Lighting Director	
Technical Director	
Audio Director	
Set Designer	
Set Construction	
Videotape Operator	
Camera #1	James
Camera #2	Larry
Camera #3	Michelle

Gaffer	_____
Grip	Tony
	Sharon
Makeup	_____
Hair	_____
Wardrobe	_____
Production Assistant	Marie
Asst Camera #1	Tootsie
Asst Camera #2	James
Asst Camera #3	Reed
Go-fer	Bobby

At this point, we believe this will suit our needs. We have decided to keep the crew as small as possible so as not to intrude on the event and, therefore, *you* will be responsible for not only the director's duties but those of the lighting director, the technical director, and the audio director. The shoot does not require a videotape operator, set designer, set construction crew, makeup, hair, or wardrobe.

CONCLUSION

The equipment and crew lists will help us plan a video within our capabilities. The two factors of what equipment is available and what crew we have to run it are themselves as important as the creation of a good idea for the video. A particular idea may have the potential for the greatest video ever. The script may be the best you have ever seen. The effects may be the most dramatic. And the characaters may be the most memorable. But if you cannot shoot it, that is, if you do not have the equipment and crew available to shoot it, it is just a bunch of words on paper.

If you have an unlimited budget, you do not have be too concerned with equipment and crew. You will be able to buy or rent whatever you need in equipment and hire whoever you need to run it. But if you are limited by what equipment and crew you have access to, then the equipment and crew lists become items you should research *before* you develop the show itself.

Coming up next: *The Format* and the development of the idea continues.

TO DO

1. Complete your Equipment List.
2. Complete your Crew List.
3. Check both the Equipment and Crew List off the Master Checklist.

CHAPTER SIX

THE FORMAT

Materials needed for this chapter:

Completed
SHOW OUTLINE

Still Working On
MASTER CHECKLIST

Blank Forms
FORMAT
SCENARIO

We have a concept. We have a tentative taping date. We have an idea of what equipment and crew we might have available. We are now ready to get serious about this video.

You could tape the video with the outline only, assuming it was a well thought-out outline. However, it is advisable to take your thinking process at least one step further and write a format. The format, as demonstrated in Figure 6–1, includes:

1. the pictures we see
2. the action
3. the sound.

Figure 6–1 The format is a narrative interpretation of the pictures you see, the action taking place, and the music or sound being heard.

The *format*, you will remember, is an extension of the outline, a step-by-step narrative of what is going to happen in the video. It does not designate exact visuals or dialogue. It is the framework, not the whole house. A script, on the other hand, *is* the house. It is everything that is going to happen, in the order that it will happen.

Formats are not hard. Basically you take the information from the outline and you add to it to develop a more complete picture of what you will see and hear in the video. You might say a format is simply an outline in narrative form.

In the professional television world, formats are sometimes called *story treatments* and are written in the style of the one that follows, written for a new game show for one of the major networks.

First, the outline for the show:

SUBJECT   Treasure Hunt Game Show
PURPOSE   Broadcast Network Television, Saturday Morning
ROLES   Host, Prize Girl, Voice-Over Announcer, 3 Studio Contestants and 3 Field Contestants
THE PLOT   Contestants search in teams of two for treasure. Each team is given a specific time limit and the team finding the most treasure within that time limit wins all the treasure they find.
THE CLIMAX   The winning team gets to search for cash money.
THE CLOSE   Contestants congratulating each other on their winnings.
THE CREDITS   Roll over closing video sequence
THE OPEN   Open on black. Start music with audio visualizer creating visual effects popping out of the black. Dolly camera to full shot of set that is backlit only. Lights are not up full. Laser light shoots out of one part of set writing the name of the show on the other part. Contestants enter, turn up lights on set, and then the Host enters to begin the show.
THE PICTURES   Closeups intercut with establishing wide shots.

THE MUSIC   Special composed especially for the show.

THE SOUND EFFECTS   Various as needed to signal the beginning of a race, the end of a race, error, etc.

THE LOCATIONS   #1. Studio. homebase for the Host and 3 contestants. A high-tech environment with each contestant having his own working console which includes a television set and various buttons and bars. Also a bank of television monitors which will be the contact with the field location; #2. Field. 3 contestants team up with the 3 studio contestants to search three different locations in the field. These locations will change from week to week. The studio location will remain constant.

THE SPECIAL ELEMENTS   Headsets for all contestants. These headsets will become the communication link between the field contestant and the studio contestant. Camera in the field with the video fed back to the studio so that the studio contestant can both see and hear the field contestant, to better instruct the contestant where to go to find the treasure. Treasure maps for each location given to the studio contestant. Various stop/start clocks for each of the studio contestants. Special uniform attire for each contestant team so they can easily be identified as the Blue Team, the Orange Team, and the Yellow Team.

THE LENGTH   Half-hour broadcast, once weekly

Now from this outline, we write this format:

OPEN on black with special visual effects full frame and shooting out of the blackness, pulsing with the beat of the music. Studio contestants enter to a set that is backlit (outlined) only. Contestants push buttons and pull fader bars on their consoles to bring up special visual effect on monitor wall, and to bring up lights and special effects in the contestants' set unit. Music builds to full and laser light shoots out of the contestants' set to the monitor wall, writing the name of the show in the blackness above the monitors. Host flips in as music and effect reach their max.

Host intros the show and today's remote location, then throws to commercial.

COMMERCIAL BREAK

ACT I.   Host intros contestants and welcomes viewers to the treasure hunt. Studio contestants punch up their team members at remote locations by depressing buttons on their consoles. Field contestants introduce themselves.

Host explains rules of the treasure hunt.

Field contestants intro the locations they will be searching—name and one-shot visual—a more detailed view of each location will precede each search.

First team to search is chosen. Blue team goes first. Blue team gets ready to search. Field contestant at starting point. Location is previewed with hiding places noted. Studio contestant turns over map and sets clock. Hunt begins with studio contestant starting clock.

Treasure Hunt #1—Blue Team—3 minutes

Video during the search is split screen, over the shoulder of the studio contestant who is working with the map and the field contestant racing through the location searching for the treasure.

Time ends. Prizes are awarded.

Host throws to break.

COMMERCIAL BREAK

The foregoing sample format continued, of course, through Act II, Act III, and the close but we will not go further. However, if we have interested you in the show, check out the script at the back of the book. A condensed version of the show, called *Finders Keepers,* is there. Maybe you'd like to take a shot at taping your own version. At any rate, back to formats for now.

You can see how with the example given here we have taken an idea for a treasure hunt game show and developed it into a show. We could take this format and shoot the show but we would be leaving a lot to chance. For example, what will the host say when he enters for the first time? What questions will he ask the contestants? How exactly will the game be played? When does it start? When does it end? There are a lot of unanswered questions that can be answered only by a script. Since television does not like surprises, virtually everything you see on broadcast TV is scripted. Even the game shows work from a basic script. But scripts later; formats now.

HOW DO YOU WRITE A FORMAT?

To make it easier to write a FORMAT for our video, we have developed two forms shown in Figures 6–2 and 6–3 that break down the format into specific parts. Actually, these forms are more complete than the narrative format. They give you more specific information and thus will be easier to shoot from if you do not elect to go further and write a script.

Show Title _____

Date _____

FORMAT FORM

FORMAT NO. _____   (Same as scenario no.)

    DESCRIPTION: _____

    _____

    SCENE LOCATION: _____

        INTERIORS: Yes _____   No _____

        Where? _____

        EXTERIORS: Yes _____   No _____

        Where? _____

SPECIAL

        MUSIC: Yes _____   No _____

        What? _____

        SOUND: Yes _____   No _____

        What? _____

        LIGHTING: Yes _____   No _____

        What? _____

        SET: Yes _____   No _____

        What? _____

        PROPS: Yes _____   No _____

        What? _____

        SPECIAL ELEMENTS: Yes _____   No _____

        What? _____

        _____

        _____

Figure 6–2 Format form.

SHOTS TO TAPE:

Scene No.

————— 1. _____

_____

————— 2. _____

_____

————— 3. _____

_____

————— 4. _____

_____

————— 5. _____

_____

————— 6. _____

_____

TAPING DATE _____

Figure 6–2 (*cont.*)

The Scenario

First, the *format scenario*. This is a list of the sequence of events in the order you see them happening (Figure 6–3). It is a tool to help you visualize the complete video from beginning to end. Try to visualize each event and how you see it framed. Write these events down on the scenario. Following is the way we see our wedding video right now:

FORMAT SCENARIO

Sequence of Events

1. Mary and John meet and fall in love, shown in a series of snapshots.
2. John asks Mary to marry him.
3. Mary buys her wedding dress.
4. Mary's little brother talks about his sister.
5. Mary talks about John.

Show Title _____

Date _____

SCENARIO

(Add to this as needed)

Location   Sequence of Events
_____   _____

_____   1. _____
                                        _____

_____   2. _____
                                        _____

_____   3. _____
                                        _____

_____   4. _____
                                        _____

_____   5. _____
                                        _____

_____   6. _____
                                        _____

_____   7. _____
                                        _____

_____   8. _____
                                        _____

_____   9. _____
                                        _____

_____   10. _____
                                        _____

_____   11. _____
                                        _____

Figure 6–3 Scenario form.

_____ 12. _____
                             _____
_____ 13. _____
                             _____
_____ 14. _____
                             _____
_____ 15. _____
                             _____
_____ 16. _____
                             _____

Figure 6–3 (*cont.*)

6. John talks about Mary.
7. Mary dresses for the wedding.
8. John dresses for the wedding.
9. Mary's best friend talks about Mary and John.
10. The wedding ceremony.
11. The reception.
12. They leave on their honeymoon.

Now that we have the video in our heads, let's break down this visualization into specific locations. Take each of the events and designate the location where each sequence will be taped. Do not designate specific areas at a location but just the main location site; such as Mary's house, not Mary's bedroom. For example:

FORMAT SCENARIO

Sequence of Events

LOCATION	
Studio	1. Mary and John meet and fall in love shown in a series of snapshots.
Park	2. John asks Mary to marry him.
Store	3. Mary buys her wedding dress.
Mary's house	4. Mary's little brother talks about his sister.
Mary's house	5. Mary talks about John.
Mary's house	6. John talks about Mary.
Mary's house	7. Mary dresses for the wedding.

Church	8. John dresses for the wedding.
Mary's house	9. Mary's best friend talks about Mary and John.
Church	10. The wedding ceremony.
Reception	11. The reception.
Reception	12. They leave on their honeymoon.

This is our format scenario from which we will now write the format.

The Format

Taking it a step further, we now want to prepare a format form for every event on the scenario. Thus, we would prepare for this particular sequence a total of twelve format forms since we have twelve events.

You will notice that included on the format form (Figure 6-2) are the pictures, sound, and music as well as any special needs such as set, props, lighting, and special elements. We will also note taping dates for each location.

While you are completing your format, remember to refer to your outline. It is your guide. Also, include as much specific detail in your format as you can, such as the name of a song or a specific shot.

You may have more than one format form for each location. For example, we will have five format sheets for Mary's house, one for each sequence of events on the scenario. This does not mean, however, that we have to roll tape at that location on different days. On the contrary, we should schedule all taping at a location on the same day, if possible. It does not matter that we tape the show out of sequence in terms of the way it will be seen when finished. The pieces will be put together in their correct order at the editing session. That, after all, is what the edit is for, to bring it all together into a workable video. We will deal with specific taping dates later in this chapter.

The format form provided here is in essence a miniscript. It does everything but lock in the words. While you think about it in reference to the video you have in mind, let's put it to work on our wedding video. We will complete all of the form except the taping date. We will decide this after all format forms are complete.

FORMAT # 1

DESCRIPTION John and Mary meet and fall in love, shown in series of

snapshots.

SCENE LOCATION: Studio

INTERIORS: Yes   X   No

Where?   Director's house

EXTERIORS: Yes   No   X

Where? _____

SPECIAL

    MUSIC: Yes _____ No ___X___ What? _____

    SOUND: Yes _____ No ___X___ What? _____

    LIGHTING: Yes ___X___ No _____ What? LAMP _____

    SET: Yes _____ No ___X___ What? _____

    PROPS: Yes _____ No ___X___ What? _____

    _____

    _____

    SPECIAL ELEMENTS. Yes ___X___ No _____ What? SNAPSHOTS _____

    _____

    _____

    _____

SHOTS TO TAPE:

    1–1 Various snapshots

TAPING DATE _____

    Since we are dealing in this scene with snapshots, we are planning to record this *before* the day of the wedding. That way, we will have it on tape and ready and not have to worry about taping it after the wedding when the edit will be our main focus. We have chosen a studio as our location. This can be anywhere we like as long as the site gives us the room to get the shots needed. We plan, at this point, to tape these snapshots in our own home.

    Note that we have numbered our shots to tape, in this case only one, by first using the scene number and then a consecutive second number.

FORMAT # 2

    DESCRIPTION John asks Mary to marry him. _____

    SCENE LOCATION: Park _____

    INTERIORS: Yes _____ No ___X___

    Where? _____

        EXTERIORS: Yes ___X___   No _____

        Where? By lake _____

SPECIAL

        MUSIC: Yes ___X___   No _____   What? "The Way We Were" _____

        SOUND: Yes _____   No _X_   What? _____

        LIGHTING: Yes _____   No _X_   What? _____

        SET: Yes _____   No _X_   What? _____

        PROPS: Yes _____   No _X_   What? _____

        SPECIAL ELEMENTS: Yes _____   No _X_   What? _____

SHOTS TO TAPE:

        2–1 Wide shot John and Mary walking in park _____

        2–2 2 shot as John pops the question _____

        2–3 CU Mary's answer _____

        2–4 CU 2 shot as they kiss _____

        2–5 Wide shot park _____

TAPING DATE: _____

Again, note the shots to tape are numbered first with the scene number, in the above case #2, followed by a consecutive number, 1 through 5. This numbering system will be valuable later when we make our shot lists.

And the format continues:

FORMAT # __3__

        DESCRIPTION: Mary buys her wedding dress. _____

        SCENE LOCATION: Store _____

        INTERIORS: Yes ___X___   No _____

        Where? Wedding department _____

        EXTERIORS: Yes ___X___   No _____

        Where? Front of store _____

SPECIAL

      MUSIC: Yes ___X___   No _____   What? "The Way We Were" _____

      SOUND: Yes _____   No ___X___   What? _____

      LIGHTING: Yes ___X___   No _____   What? Lamp to fill in _____

                   shadows _____

      SET: Yes _____   No ___X___   What? _____

      PROPS: Yes _____   No ___X___   What? _____

      SPECIAL ELEMENTS: Yes _____   No ___X___   What? _____

SHOTS TO TAPE:

        3–1 Exterior shot with push to window _____

        3–2 Medium shot Mary looking at racks of dresses

        3–3 CU dress chosen _____

        3–4 Pull from ECU of dress detail to reveal Mary in dress, her

           Mother looking on smiling _____

TAPING DATE: _____

    And one more:

FORMAT # ___4___

      DESCRIPTION: Mary's little brother talks about his sister _____

      SCENE LOCATION: Mary's house _____

      INTERIORS: Yes _____   No ___X___

      Where? _____

      EXTERIORS: Yes ___X___   No _____

      Where? Treehouse behind main house _____

SPECIAL

      MUSIC: Yes _____   No ___X___   What? _____

      SOUND: Yes _____   No ___X___   What? _____

LIGHTING: Yes ___X___   No _____   What? _Lamp or camera_

            _light_____

SET: Yes _____   No ___X___   What? _____

PROPS: Yes _____   No ___X___   What? _____

SPECIAL ELEMENTS: Yes _____   No ___X___   What? _____

SHOTS TO TAPE:

       4—1 Wide shot treehouse as little brother climbs up _____

       4—2 CUs of little brother as he talks _____

TAPING DATE: _____

We will not continue any further with the format for the wedding video. You get the idea. Once all format forms are complete, you will have a complete picture, from beginning to end, of what will be happening in your video and where it will happen.

HOW DO YOU USE THE FORMAT?

The format has allowed you to "see" the video better. As a matter of fact, it may be complete enough to tape from, especially if you have a spontaneous situation that requires no written dialogue. What you do with the format now depends on whether or not you plan to write a script.

If You Are Writing a Script

If your video has any dialogue that must be written, you will want to write a script or at least a short script. If you do this, you will use the format as the base for the writing of that script or short script. All the information on the format will be incorporated into that formal written version of the show. Even shot numbers can stay the same, if you like. Or you may elect to use the shot information only and renumber the shots on the script. Do whichever is easier for you. The number on the shot is not important. It is important only that it be numbered.

You may find that as you write the script, you want to change things you decided in the format. You may even decide to eliminate a scene entirely or to add another totally new one. That is fine. The format is the base, not the Bible for the script. Use it as that.

If You Are Taping from the Format

If you elect to tape from the format and not go further with the show development process, which includes the writing of a script and/or storyboards, you will now want to do the following:

1.   Divide your completed format forms into locations, one stack for every location. Our twelve format forms for the wedding video, for example, would be divided with five forms in the stack for Mary's house, two in the reception stack, two in the church stack, one in the store, one in the studio, and one in the park.

2.   Now look at every stack individually to decide whether all the shots at any one location can be taped on the same day. For example, in the stack for Mary's house, we have a format form for the following scenes:

Scene 4	Mary's little brother talks about his sister
Scene 5	Mary talks about John
Scene 6	John talks about Mary
Scene 7	Mary dresses for the wedding
Scene 9	Mary's best friend talks about Mary and John

With the exception of Scene 7, all these scenes can be taped prior to the wedding day. Since our goal is to get as much on tape prior to the actual day when the pressure will be at its greatest, it makes sense to schedule a day prior to that day to tape Scenes 4, 5, 6, and 9. Let's say we select November 1 and 2, two days to be on the safe side. Weather, after all, could force us to change the taping day of our exterior shot with Mary's little brother from one day to another. The taping date for Scene 7 is November 15, the day of the wedding.

If you have time, you can schedule more than one location per day, but remember that you must have time to set up properly at each location before you can roll tape. We could, for example, plan to tape at the park on November 2 after we wrap at Mary's house.

Continue this process with all your format forms until all have a taping date written in.

3.   Shuffle the format forms again, this time putting them in sequential order according to the taping date.

4.   Now take each taping date separately and decide in what order the scenes will be shot. What will be shot first? What last? Is there a natural progression through the location that will make the taping go easier? Will it be simpler in terms of the setup to move from, for example, the treehouse to Mary's bedroom to the living room or will another sequence be better? Make this decision based on the layout of the location.

You now have your format forms stacked by taping date as well as shot order. From this stack you can create a taping schedule (see Figure 2–3) for the entire shoot. You could even tape from these forms or you can take the idea one step further and write a script, which we will do in the next chapter.

CONCLUSION

The format scenario breaks down the concept into specific scenes. The format breaks these scenes down into specific shots noting all the visual and audio elements that will be used to create the desired effect with that scene. The format numbers all shots, which will make it easy to create a shot list should you decide not to write a formalized script. Shots are numbered by scene number and then in consecutive order at the scene.

A format may replace a script if the show is spontaneous, not requiring exact dialogue, or a format may be just another show development tool forming the base for the writing of the script. You make the final decision of whether to use the format as your final shooting form or to go on.

Coming up next: *The Script* . . . how to write one.

TO DO

1. Complete a Scenario.
2. Complete a Format form for every sequence of actions at every location.
3. Check off the Scenario and the Format on your Master Checklist.

CHAPTER SEVEN

THE SCRIPT

Materials needed for this chapter:

Completed

SHOW OUTLINE
FORMAT

Still Working On

MASTER CHECKLIST

The script is the ultimate plan for the video. It is, as shown in Figure 7–1, detailed specifics of everything in the show, up to and including:

1. *Exact* dialogue or a general idea of what will be said in a live or impromptu situation or interview
2. Specifically what visuals will be seen and where they will be seen
3. Any effects in lighting, pictures, sound, or music, whether taped or done in the edit session
4. Specifically when sounds will be used or music played
5. Any titles, graphics, or words to appear on screen.

The script may show transitions from one scene to another as well as how you want a shot framed. It is your blueprint for the video (Figure 7–2). If it is written well, you only need follow it to create exactly what you had in mind. Knowing exactly what video and audio will be recorded and how it will be recorded will give

Figure 7–1 The parts of a script.

you more control of the show. For this reason, we strongly suggest that you consider writing one. You should not, however, consider any script to be sealed in concrete. You should always maintain some flexibility. Some of the best shots and the best dialogue and the best effects have been created during taping or editing. You cannot

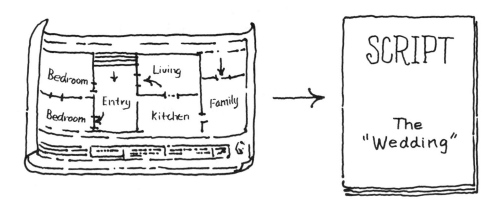

Figure 7–2 The script is the blueprint for the show.

Figure 7–3 A script helps you to visualize what the show will look like.

predict everything that will happen during taping, especially when you are dealing with amateur talent in a live situation. You should, therefore, always be prepared to incorporate new ideas that will make the video better.

A script will help you visualize the show, as shown in Figure 7–3. It will direct your mind to the fine details and force you to see how they will look, hear how they will sound and know how the pieces will all fit together. A script will save you time in the taping because it forces you to make decisions, to take control rather than just randomly roll tape. It will allow you to detail every element so that you can make certain that you tape everything and that you take best advantage of all the elements, including equipment, crew, and cast available to you. It will make it possible for you to number every shot and plan the taping sequence in detail. It will save you time in the edit because it will give you a specific plan to follow.

HOW DO YOU WRITE A SCRIPT?

You may think that a script is hard, even impossible to write. Not so. We won't say that it is easy to write a script because for some perhaps it is not. But we will say that there is no one who cannot write one.

Everything in the script falls into one of two categories, video or audio. As we have already discussed in Chapter 1, video is everything you see; audio is everything you hear. The video and audio in a script will have a specific place in the design of the script and each will be written in a certain form. All of this is to make it easy for you to distinguish, at a glance, one from the other. Before we deal with the placement of these two elements, let's discover exactly what is included in each.

The Video

The video is, as we have said, everything you see. It includes, as seen in Figure 7–4, the scene you are recording, titles, graphics, any prerecorded video or film footage, and any visual effects you plan to incorporate into the show. All video on the script is always written in all capitals. If it is a Video Special Effect, the description of the video is preceded by VIDEO EFX: or just VFX:.

Figure 7–4   Video includes the scene you are recording: titles, graphics, any old video, film footage, and any visual effects. Everything you see.

The video description on the script may also include transitions from one scene to another. These transitions might include FADE IN or OUT, CUT, DISSOLVE, WIPE IN or OUT and any number of special effect transitions. Your transition abilities will, of course, depend on the capabilities of your equipment. As we have mentioned before, you should make a point of discovering just what your equipment can do before scripting so that you do not write in a transition that is not possible to do. Most standard camcorders can FADE IN, FADE OUT, CUT, and perhaps do a simple WIPE. Most do not have special effect or dissolve capabilities.

The video description on the script may also include shot framing. Shot framing commands include EXTREME CLOSEUP (ECU), CLOSEUP (CU), MEDIUM SHOT (MS), MEDIUM WIDE SHOT (MWS), and WIDE SHOT (WS). If there are people in the shot, you can fine-tune these commands with TWO SHOT (2S), THREE SHOT (3S), FOUR SHOT (4S), CROWD SHOT (CS), all numbers noting the number of people in the shot. These commands are usually abbreviated as indicated in the parentheses above.

In addition, some camera calls may also be included in the script, such as. PAN RIGHT or LEFT, TRUCK RIGHT or LEFT, TILT UP or DOWN, ZOOM IN or OUT or RACK FOCUS. Even if you allow your camera operators to frame their own shots, it is advisable to write in something to at least give them an idea of what you have in mind. In a professional script, only basic shot framing and camera calls are written in by the writer. It is up to the director to add to these to create a show that is representative of his or her own style. The director creates the show just as much as the writer does.

Putting all the elements together, the video portion of a script might read like this:

FADE IN ON MS OF JOHN AND MARY EATING WEDDING CAKE.
VIDEO EFX: CAKE FALLS IN SLOW MOTION TO THE FLOOR.

Note that the video description on the script is always written in all capital letters.

The Audio

The audio includes everything you hear. There are three types of audio in a script, as shown in Figure 7–5:

1. Sound
2. Music
3. Dialogue

Figure 7–5   Audio is sound, music, and dialogue. Everything you hear.

Sound/Music.   Sounds and music are always written in all capitals and underlined. Specific music and sounds are noted if the selection has already been made. If not, a description is appropriate.

MUSIC: ROMANTIC SONG

SOUND: SCREECHING OF CAR TIRES

An audio effect is written   SOUND EFX: or just   SFX: and followed by a description of the effect.

SOUND EFX: SPACE SHIPS FIRING LASERS

The duration or way the audio is brought into or out of the scene may also be indicated where appropriate.

MUSIC: FADE UP ON "THE WAY WE WERE"

You can include an indication of the intensity of the sound or music with:

MUSIC: UP FULL

This means that the music is up as loud as it is ever going to be in the show. You might want to indicate music as background with:

MUSIC: UNDER

Normal sounds are noted by a description of the sound desired.

SOUND: TRAIN WHISTLE

If the sound or music has a natural end and is short in duration, such as the train whistle, it is not necessary to indicate that the sound or music has stopped. On the other hand, if the music or sound continues for a time, then you must indicate on the script when you want it to stop. Otherwise, it is assumed that you want it to continue.

MUSIC: "THE WAY WE WERE" FADES OUT

You can cut or fade out audio as already suggested. A cut is an abrupt stop; a fade is a gradual going away. If you know the name of the music you want to use, write the name into the script. If you do not, then just write the kind of music you have in mind. The director with the assistance of the A.D. will make the final choice.

Sound and music commands then include: FADE UP, FADE OUT, CUT IN, CUT OUT, UP FULL, or UNDER.

Dialogue. Anything said by the performers is called *dialogue*. Dialogue can be delivered on camera or voice over. On the script, all dialogue is preceded by the name of the performer, underlined and in all caps. This is followed by the words that the performer will say, written in upper and lower case, as normal. Generally, the performers are asked to memorize these words and to deliver them just as they are written unless you, as the producer/director, tell them differently. If the dialogue is part of an interview, you may note the questions to be asked with the response indicated only by the word (response), in parentheses as demonstrated in Figure 7–6. The parentheses indicate that the words inside the parentheses are not to be said, but that they are stage directions for the performer; in this case, a direction to respond in a normal way to the question asked.

Other stage directions for the performer may be written before, within, or following dialogue as shown in Figure 7–7. A direction is put into the performer's dialogue in this way to make certain that the performer reacts as is intended.

SCRIPT

MARY: What is
your name?
JOHN: (Responds)

Figure 7–6 Unscripted or impromptu dialogue as it is indicated in the script.

Putting both video and audio together, the script may look like this: (See Figure 7–8 for a visualization of this script sequence.):

CUT TO RECEPTION HALL. MARY AND JOHN ARE FEEDING EACH OTHER CAKE. MARY DROPS A PIECE. CAKE HITS THE FLOOR AT JOHN AND MARY'S FEET.

SOUND EFX: CAR CRASH

<u>JOHN SMITH</u>

(Laughing) That was really great timing,
Mary. (Bends to pick up cake) I couldn't have
done it better myself.

JOHN: (enters
laughing) That
was really good
timing, Mary.

Figure 7–7 Stage direction for the talent is also written into the script.

<u>MARY SMITH</u>

(Responds)

You will note that we have not written into the script some of the visuals seen in Figure 7–8, for example the CU of John laughing, the cake falling, or John picking it up. These are shots the director would call for during the taping or we would pick them up as cutaways for insert during the edit. The script leaves most of the specifics of the shots in the hands of the director.

The basic elements included in the script then are:

<u>VIDEO—written in all capital letters</u>
      TITLES
      GRAPHICS
      SPECIAL EFFECTS (Video EFX)
      SCENE DESCRIPTION
            with sometimes an indication of:
                  TRANSITIONS
                  SHOT FRAMING

<u>AUDIO—written in all capital letters and underlined except dialogue</u>
      <u>which is written upper and lower case, as normal</u>
      SOUND
      MUSIC
      SPECIAL EFFECTS (Audio EFX)
            with sometimes an indication of:
                  DURATION
                  INTENSITY
      DIALOGUE

The Script Style

There are two different styles used in script writing. Both are acceptable. Both are professional, but generally each style is used for a certain situation. Style 1, shown in Figure 7–9, is used for broadcast commercials or shorter videos produced for nonbroadcast corporate or company use. Style 2, shown in Figure 7–10, is used for dramatic shows on Broadcast television, such as *The Love Boat* and *The Cosby Show*. Either style works and incorporates all the same elements. The difference is only in the form, not in the information. Some writers and directors prefer Style 1 while others prefer Style 2. Choose the style that you find easiest to use.

CUT TO MS
JOHN AND MARY FEED
EACH OTHER CAKE.
MARY DROPS A
PIECE.

CUT TO CU 2 SHOT
OF JOHN AND MARY.

John: That was really great

CUT TO CU OF
CAKE AS IT HITS
FLOOR.
SOUND: CAR CRASH

CUT TO CU JOHN
PICKING UP CAKE.

John: I couldn't...

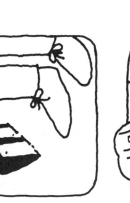

CUT TO CU OF
JOHN.

John (laughing)

CUT TO CU 2 SHOT
OF JOHN AND MARY

John: have done it
better myself.
Mary: I know.

Figure 7–8 All elements of the script as it would be written and how it might be shot.

SCRIPT

STYLE 1

VIDEO OR EFX. ALL CAPS NOT
UNDERLINED. XXXXXXXXXXXXXXX
XXXXXXXXXXXXXXX SOUND/MUSIC/
AUDIO EFX: ALL CAPS. UNDERLINED.
XXXXXXXXXXXXXXXXX

STAR'S NAME. UNDERLINED.
Dialogue. Upper/lower case.
    Xxxxxxxxxxxxxx. Xxxxxxxxxxx.
Xxxxxx. Xxxxxxxxxxxxxxxxxxxxxx
    xxxxxxxxxxxxxxxxxxxxxxxxxxxx
xxxxxxxxxxxxxxxxxxxx. Xxxxxxxxx
xxxxxxxxxxxxxxxxxxxxxxxxxxxxx.

Figure 7–9   Script Style #1.

STYLE 2

VIDEO/VIDEO EFX: ALL CAPS. NOT UNDERLINED.XXXXXXXXXXXXXXXXXXX
XXXXXXXXXXXXXXXXXXXXXXXXXXXXXXXXXXXXXXXXXXXXXXXXXXXX

SOUND/MUSIC/AUDIO EFX: ALL CAPS. UNDERLINED. XXXXXXXXXXXXXXX
XXXXXXXXXXXXXXXXXXXXXXXXXXXXXXXXXXXXXXXXXXXXXXXXXX

STAR'S NAME. UNDERLINED

Dialogue. Upper/lower case. Xxxxxxxxxxxxxx
xxxx xxxxxxxxxxx xx xxxxxxxx. Xxxxx xxx

Figure 7–10   Script Style #2.

Style 1.   Imagine a line drawn down the center of your page, dividing it in half. On the left side of the page, you write the video including the transitions, any camera direction, a description of the scene, and the action that takes place. All this is written in all capitals. This video does not go beyond the center of the page, your imaginary boundary. Also written on this side of the page is all music or sound.

The right side of the page is reserved for dialogue. You write the name of the performer in all capitals followed by dialogue in upper and lower case and any stage direction that goes along with that dialogue. If there is no dialogue for the video, the right side is blank.

This script style is easy to read since the video is always on the left half of the page and the spoken word is always on the right. There is little room for confusion between the two and you can easily check off each scene as it is recorded. A script written in this style can be found in Chapter 15, the popcorn commercial.

Style 2.   This style utilizes the entire width of the page with the video and all audio except dialogue written across the full width. Dialogue is indented from both sides of the page, placing it in the center of the page. As in Style 1, all video is written in all capital letters. Audio is written in all capital letters and underlined including sound, music, and audio effects. Dialogue is written in upper and lower case, as normal.

In Style 2 there may be room for confusion, since the video and audio of any particular scene flow together, rather than being sharply distinguished from each other as they are in Style 1. Style 2 is sometimes called film-style scripting since it was and still is the style used for film scripts.

Things You Need to Know Before You Write

Before you write a script, you need to have clear in your mind the answers to certain questions. These, as demonstrated in Figure 7-11, are:

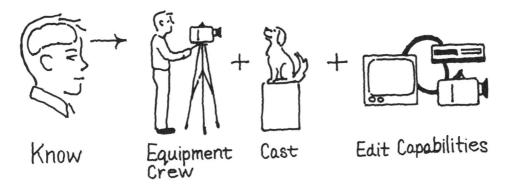

Know   Equipment   Cast   Edit Capabilities
              Crew

Figure 7-11   Before you write your script, you should have the answer to certain questions clear in your mind.

1. What equipment is available to you? You will need to know this so that you will not write a script beyond the capability of the equipment. Writing a four-camera shoot and having access to only two cameras can be frustrating, not to mention it would require extensive rethinking of the taping process. Refer to your equipment list.

2. What crew is available to you? Again, you do not want to write beyond the limitations of your crew, either in numbers of crew available or their specific skills in creating videos. Refer to your crew list.

3. What performers will you have in the video? Sometimes you can just write a video and cast the production later. Sometimes, your performers will be dictated for you, as in our wedding video. Whatever the case, know your cast, their limitations, and your own requirements. You want them all to perform at their best. Do not write beyond their capabilities. Refer to the basic cast list already on your outline.

4. What are your editing capabilities? While you might not think that this is important at the time of taping, you are wrong. How good the video is depends on not only how it is shot but how it is edited. There are many things you can do in the taping process that will make up for the lack of editing facilities if you know in advance what you will be able to do in the edit. Know your editing capabilities.

WHAT ARE SOME SPECIAL WRITING SITUATIONS?

You cannot always just sit down and write a script. As a matter of fact, you will probably find yourself continually dealing with special situations that require special writing. We hope to help you deal with some of those special situations with the following suggestions.

Writing for the Amateur

Everyone thinks they are a star *until* the camera starts rolling. Then they may become self-conscious, nervous, and even panicky. They will mainly be terrified that they might come off looking foolish. Ironically, it is just that fear that will make them look foolish. If you are dealing with amateurs, your first job is to get them to relax. A tightly written script may not be the answer. While such a script will assure you that the show will include what you want said, reading or memorizing someone else's words may cause your amateur talent to completely freeze up.

Generally, writing words for amateurs is not a good idea. It is best to let the amateur improvise and then you can edit to the best delivery. A good script only makes a good video when you have good talent to deliver the lines. When you have amateurs, it is wise to use the script to point them in the right direction. Never marry them to the words. You may not be happy with the results.

Improvised Dialogue

Improvised dialogue includes impromptu or spontaneous situations such as interviews or live events. In a sense, you write the script for improvised dialogue when you plan the edit. Not knowing what will be said in a situation like this, you can only roll tape and pick and choose later from what is recorded.

You may find that your video includes some of this kind of impromptu dialogue. Don't fight it by trying to structure it. You can write questions if you like or make suggestions but it is best to just let these live situations follow their natural course, keep as detailed notes as you can, and make it work in the edit session.

Writing from Existing Material

You may wish to create a video from existing tape. In other words, you may have been accumulating shots of the baby and have decided to create a video of *Baby's First Year*. Would you write a script for such a video? The answer is a resounding, *yes*. Writing a script from existing material can be easy and it will give more direction toward the desired outcome. Without a script, you may find yourself wandering through the tape time after time, unable to make any decisions because you just cannot quite remember all the good stuff that is there. Also, without a script, you will be simply editing pieces together with no audio continuity. Perhaps a voice-over will be needed to explain the different shots or your feelings about the subject. A script will allow you to bring these bits and pieces together into one cohesive, exciting video. You would begin such a project just like it was a new video, from the beginning outline to the format to the script. It will become obvious that certain steps can be bypassed along the way, particularly the planning related to the taping session if you already have absolutely everything on tape, but the basic process is still the same if you want to create a good video.

IS THERE A SHORT WAY TO WRITE A SCRIPT?

Actually there is a short way to create a script for your video. You have already completed the basics of this *short script* with the format you made in the last chapter. You will remember that the format is extremely detailed, including location, all audio and video elements, and even shots designated by number. There is really not much more to add to this to make it a viable short script. Basically, the only thing missing is dialogue. Specifically, what will be said and who will say it? This would include actual words that will be said or a structured guideline for an impromptu delivery. That is basically all you need to add to make the format into a complete short script. It will, of course, be written in a different form from the way a script is usually written, but all the information will be there and it will work fine for a video that does not require a full-blown script.

As a matter of fact, we recommend this short script for a video that requires little structured dialogue, like the wedding video. We do not recommend it for a video that requires that all the dialogue be written. It would be best, in that case, to write the script in one of the usual script styles. It would be too confusing to try to figure out where and when all this extensive dialogue was supposed to be delivered without the indication of video and audio in its proper place on the script form. But a short script is just fine for the wedding where we may want to write some dialogue but certainly not much. We might, for example, choose to write some dialogue based on a previous interview with the bride's mother for scene 11, the reception, as follows:

<u>Bride's Mother</u>

My Mary is such a sweet girl. I know she will make a great wife and a great mother. They're going to live in the Caribbean, you know. Isn't that nice.

We do not really consider this structured dialogue since certainly we would not ask Mary's mother to memorize or read it. Rather we will tell her that those are the words we liked best in the preinterview and ask her if she can limit herself to those. It is a subtle way to get the amateur to deliver the lines we want without intimidating them with a formal script. They feel more comfortable because the words are theirs, not yours.

If the situation does not call for the formal script but does need some development beyond the format, try the short script concept. It may be all you need.

CONCLUSION

A video may be totally scripted or it may be partially scripted. The decision of whether you need formalized dialogue and more complete shot descriptions is answered by the video itself and your own interpretation of that video. A formal script includes all video and audio for the show. But even as complete as a script may be, it still requires some interpretation by the director, for it is she or he who decides specifically how a shot will be framed, how long that shot will stay on, and from what POV a scene will be shot. Ultimately, the video is the director's show, not the writer's.

Here's a hint for when you are writing a script: it takes approximately two typed pages or four handwritten pages to create one minute of videotape. Keep this in mind when you are writing your script. It will help you to determine approximately how long your video will be. We strongly suggest that you do not make a video longer than twenty minutes, the best length for keeping interest, as noted previously. If you keep yourself to this length, you will be able to spend more time on the product, hopefully creating a very tightly edited, exciting video that your family and friends will be willing to view over and over again.

Coming up next: *The Storyboard* . . . how to create pictures of the show without rolling tape.

TO DO

1. Complete your Script.
2. Check the Script off your Master Checklist.

CHAPTER EIGHT

THE STORYBOARD

Materials needed for this chapter:

Completed

SHOW OUTLINE
FORMAT or SCRIPT

Still Working On

MASTER CHECKLIST

Blank Forms

STORYBOARD
STORYBOARD GRAPHICS

Storyboards are the visualization of the show on paper before tape is rolled. They are pictures drawn so that everyone can see certain scenes or, in some cases, the entire show. They are usually basic black and white sketches, showing only pertinent objects, people, and action, but they can be full color pictures drawn by a professional artist. Both styles are shown in Figure 8–1.

Whatever they look like artistically, the objective remains the same: to create a visual representation of the show so everyone will see it the same way.

Storyboards are generally thought of as only pictures. And in some cases, that may be all they are, *but* just a picture storyboard is the exception, not the rule. Cer-

Figure 8–1 Storyboards can be simple drawings or very elaborate finished drawings.

tainly the pictures are the number-one ingredient and you will not see a storyboard without pictures. However, a storyboard, in addition to pictures, is also likely to include (1) the words, (2) the music, and (3) any sound or special effects.

It can be extremely detailed including a drawing for every camera angle on a particular scene as we saw in the wedding cake incident, shown in Figure 8–2, now in storyboard form. Or the storyboard can be more basic, including only the visual showing the major camera angle at a particular location.

A storyboard can be as complex or as simple as you want or need it to be. Think of these drawings as a visual aid showing your crew and talent exactly what you see and, perhaps, hear in the video. Storyboards are like a cartoon strip of the show with you as the artist.

Storyboards are common in the professional video field, particularly in the taping of commercials. Commercial storyboards are usually quite complete, as you can see from the one shown in Figure 8–3. Storyboards have also been used by some of the most successful movie producers. Steven Spielberg uses them, particularly extensive ones for action scenes such as those in *Indiana Jones and The Temple of Doom,* and Alfred Hitchcock drew his own shot-by-shot storyboards for all his movies.

Some movie producers are even creating video storyboards. That is, they are taping on video *before* they begin filming. This is a rather new technique of filmmaking and we will discuss it later in this chapter.

Storyboards are recognized as the perfect way to show the crew what you, the producer/director, are seeing. It is a welcome replacement to trying to describe in words the pictures that you are seeing in your head. Your format or script could be totally reduced to these drawings. A complete storyboard could replace the format or script entirely, eliminating the need for these formal written versions. Do not take this the wrong way. We are not suggesting that you do not write a format or script. Rather, we are suggesting that perhaps once you have written one or both of these that you may want to create a complete storyboard and file the format or script away to be used as reference. It is just another alternative you have.

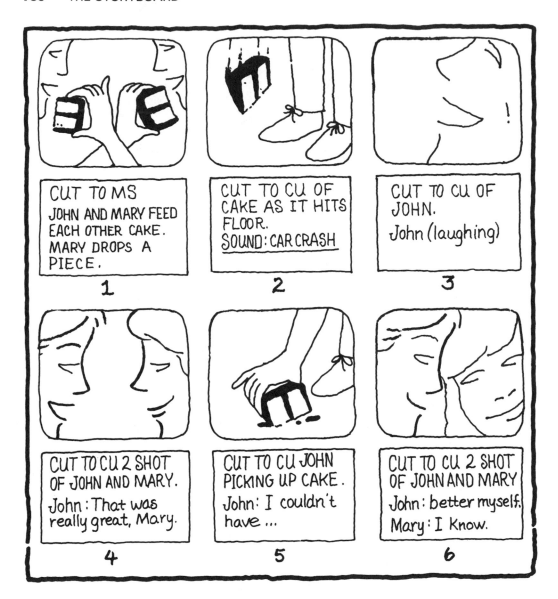

Figure 8–2 The falling cake sequence put into storyboard form.

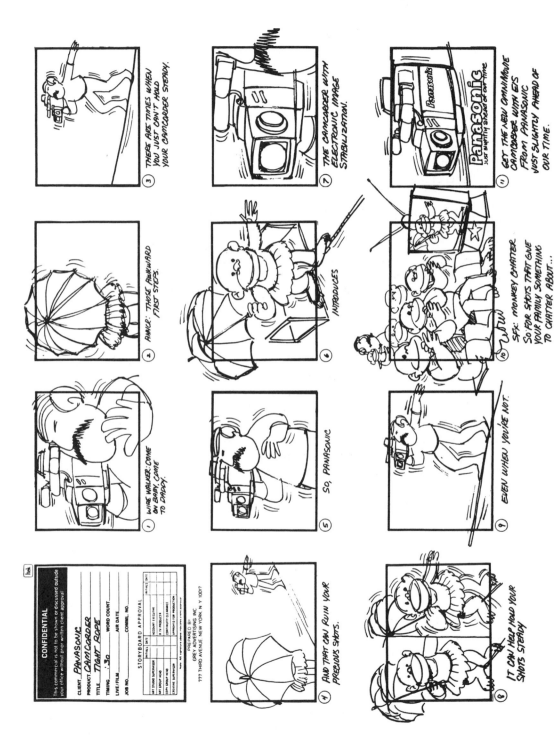

Figure 8–3 A storyboard for a TV commercial.

© Panasonic Company 1988 & 1989   **Storyboard courtesy of Panasonic Company.**

WHAT ARE THE ELEMENTS OF A
STORYBOARD?

Storyboards can include:

> Pictures
> Dialogue
> Music
> Sound
> Special effects

And, as on the script, you may note transitions and camera framing of shots as well as duration and intensity of music or sound. All of these elements will not always be included. The one exception is the pictures. They will always be in every storyboard. As a matter of fact, of the two boxes on a storyboard form, the upper box is *always* reserved for the drawing. Each individual storyboard form is called a *panel* and a complete storyboard is usually made up of a series of panels.

The Upper Box

The upper box, the picture, is the prime ingredient of the storyboard. The picture or visualization of the shot is, after all, the basic reason for creating the storyboard, to graphically show everyone what you see. Traditionally, the picture is drawn inside a graphic that looks like a TV screen so you can get TV perspective on the shot.

The picture you draw should be detailed enough to convey what you are seeing. It does not need to include every single thing that will be in the shot, only those pieces that are critical to the action. You may be drawing the picture to indicate how you want a shot framed for the camera operators or you may be drawing it to show exactly where you want certain props, furniture, or people placed. Whatever your reason, make certain to include in the picture the detail you need to convey the idea.

The Lower Box

The bottom portion of the storyboard, below the picture, is for written information including music, sound, transitions, camera framing, a description of the action, and any dialogue. This information is written in script style, observing all those rules regarding capital letters and underlining. These written elements help to take the picture beyond a mere visualization, adding action and audio, and are usually written in the following order:

Music.   This is usually the first thing noted in the larger portion of the form, below the picture. If there is no music, do not write in anything. If the music is continuing from a previous panel, you do not need to note it again. The assumption is that the music continues from one panel to the next unless you indicate that it has stopped, for example, <u>MUSIC: FADES OUT</u>. As in the script, any music indication is written in all caps and underlined.

Sound. The sound follows the music or takes its place if there is no music indication. If there is no sound or it is continuing from a previous panel, do not note sound at all. Note the end of sound only if it is a long continuing sound. And again, as in the script, sound is written in all caps and underlined.

Transition. You may want to indicate the transition to the shot shown in the storyboard, such as FADE, DISSOLVE, CUT. If this is included, it will be the first word or two in the description of the video written below the picture. A transition notation is written in all capital letters.

Camera Framing. What is the framing on the shot shown in the storyboard picture? ECU, CU, MS, MWS, WS, 2S? You may or may not include this based on how detailed the picture is. If you do include it, it is written in all capitals following the transition and prior to the description of the action.

Description of Action. This follows the transition notation and/or the camera framing. Since you have drawn the picture, you need only indicate what movement or action will take place in the scene, not the composition of the scene itself. This, too, is written in all caps.

Dialogue. The words that will be said during this visual are also written in the lower portion of the storyboard. The words should be preceded by the name of the performer delivering them. If the dialogue is continuing from a previous panel, you do not need to note the name of the performer again, but just continue the dialogue in a normal manner. Dialogue is written in upper and lower case, as normal.

Those are the main ingredients of the storyboard. Every panel includes the picture in the upper box and music, sound, transitions, camera framing, a description of the action, and any dialogue written in the lower box. Begin with the picture and then go as far you as you want from there in terms of including the other elements.

HOW DO YOU MAKE A STORYBOARD?

Storyboards do not require great drawing skills. Certainly if you are making a fancy presentation to a corporate client, you will probably hire a professional artist to do your storyboards. But for most storyboards, your own artistic skills, however limited they might be, are not only sufficient but preferable. After all, it is your visualization of the shot that you want to convey.

And remember, you will be showing your storyboards to your crew and verbally going over them step-by-step so any skills you may be lacking in drawing will be made up for with this question–answer sequence.

The Drawing

To give you a little practice in preparing your own storyboards, we have included in Figure 8–4 some drawings and in Figure 8–5 a storyboard form with the written portion already there. Make a copy of Figure 8–5 now so you can use your story-boarding skills to fill in the visual portion of this storyboard sample using the written words as your guide.

Use the copy of Figure 8–4 that you already have to cut and paste your own visualization of this western shootout, supplementing with your own drawings. Or if you prefer, you can do the entire storyboard with your own drawings. This exercise will give you some experience in preparing your own. If you use the drawings provided, cut them to the framing you want, such as medium shot or a closeup.

While we have included these drawings to help you in the preparation of your storyboard, we want to suggest that you use them only as an exercise in the form for a storyboard and as a way to get your feet wet in terms of drawing your own. When you get around to the real thing, your own drawing skills, however basic, will be good enough to convey your ideas. Even stick figures are fine for storyboards. Remember you are not creating a Van Gogh. You are only creating a basic drawing to convey an idea. The final product the world will see is not your artistic ability but the great video those storyboards will become.

The Form

To create a storyboard from scratch, you need the right form. You can make copies of the one in Figure 8–6, or the ones provided in the Appendix, or you can buy storyboard forms at your local art supply store. They can be purchased in various sizes, from an upper-screen size as small as $1\frac{1}{2}$-inch (diagonal measurement) to $5\frac{1}{2}$-inch. Or if you prefer, you can create your own form. Simply take a sheet of $8\frac{1}{2}$-inch by 11-inch paper (standard typing paper size) and draw in your TV screen. When creating your own, remember to draw the TV screen to TV perspective. You will remember the ratio is 3 high to 4 wide. A TV screen drawn 3 inches high and 4 inches wide or one $1\frac{1}{2}$-inch by 2-inch would be the correct TV ratio.

Or you could use a 4-inch by 6-inch card, drawing a line through the middle of the 6-inch side dividing the card into two 3-inch by 4-inch segments. Use the top 3-inch by 4-inch segment for the video and the bottom 3-inch by 4-inch segment for the audio. Having exactly the right ratio for your storyboard is important to demonstrate framing and composition as accurately as possible.

WHAT KIND OF STORYBOARD DO YOU WANT TO CREATE?

Do you want picture only or a complete storyboard? What is right for your project? Maybe a video storyboard is more desirable? Traditionally, there has been only one kind of storyboard, the kind drawn on a sheet of paper. However, as we have al-

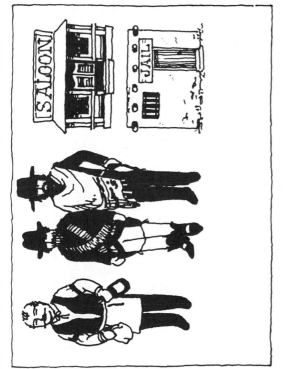

Figure 8–4 Graphics for your use in completing the western shootout storyboard.

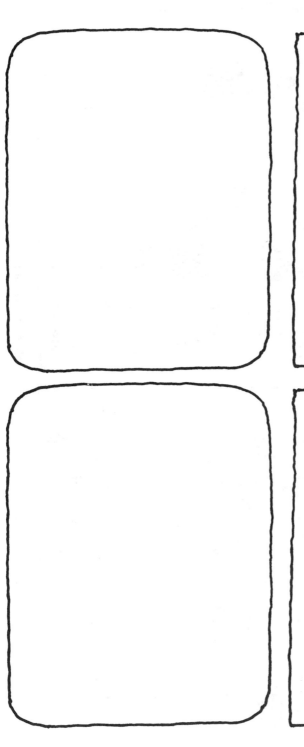

CUT TO STREET. THE MARSHAL IS WALKING TOWARD THE SALOON, FOLLOWED CLOSELY BY JERRY LEE. SLEEZE COMES OUT OF THE SALOON AND HEADS TOWARD THE MARSHAL. THEY COME WITHIN TWENTY FEET OF EACH OTHER. EACH STARING INTENTLY AT THE OTHER.

SLEEZE
What's the matter, Marshal? Afraid to take me on by yourself?

MARSHALL
I ain't afraid of you or your kind.

SLEEZE TAKES ANOTHER, DRINK FROM THE BOTTLE AND SLAMS IT DOWN ON THE COUNTER. HE BEGINS CHECKING HIS GUN TO MAKE SURE IT IS LOADED, CLICKING THE BARREL. HE PUTS IT BACK ON HIS SIDE.

SLEEZE
Barkeep, give me your shotgun.

THE BARTENDER LOOKS AT HIM FOR A SECOND, NOT MOVING.

SLEEZE
The shotgun, now.

THE BARTENDER REACHES UNDER THE BAR, PULLS OUT A SHOT GUN AND HANDS IT OVER TO SLEEZE.

Figure 8–5 The storyboard for a western shootout. Draw in the video or cut and paste using the graphics supplied in Figure 8–4.

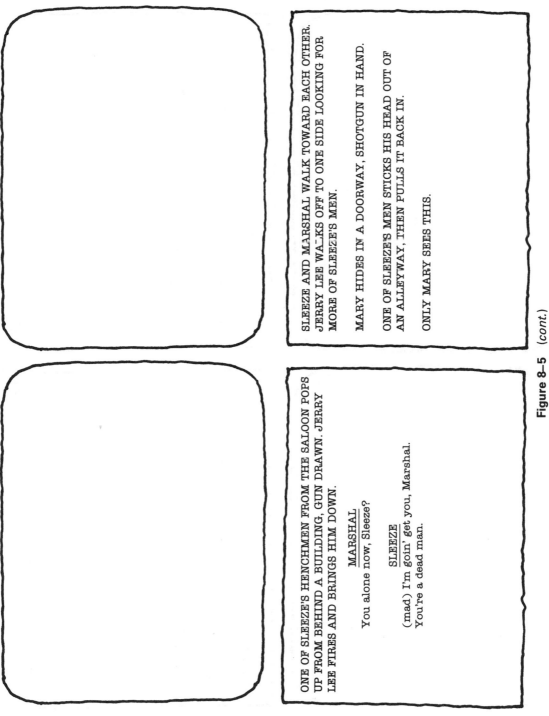

SLEEZE AND MARSHAL WALK TOWARD EACH OTHER. JERRY LEE WALKS OFF TO ONE SIDE LOOKING FOR MORE OF SLEEZE'S MEN.

MARY HIDES IN A DOORWAY, SHOTGUN IN HAND.

ONE OF SLEEZE'S MEN STICKS HIS HEAD OUT OF AN ALLEYWAY, THEN PULLS IT BACK IN.

ONLY MARY SEES THIS.

ONE OF SLEEZE'S HENCHMEN FROM THE SALOON POPS UP FROM BEHIND A BUILDING, GUN DRAWN. JERRY LEE FIRES AND BRINGS HIM DOWN.

MARSHAL
You alone now, Sleeze?

SLEEZE
(mad) I'm goin' get you, Marshal. You're a dead man.

Figure 8-5 (cont.)

MARSHAL AND SLEEZE DRAW AND FIRE.

SLEEZE GOES DOWN, MORTALLY WOUNDED. HIS GUN UNFIRED.

THE MARSHAL WALKS TOWARDS SLEEZE. MARY RUSHES OUT TO HIM.

MARY SHOOTS SLEEZE'S MAN. HE FALLS.

JERRY LEE TURNS TOWARD THE GUNSHOT AND ALMOST SHOOTS MARY.

THE MARSHAL IGNORES THIS BUT SLEEZE SEES IT.

Figure 8–5 (cont.)

Figure 8–6 Storyboard form.

ready mentioned, in addition to this conventional approach, today there is the video storyboard. Generally, we would not suggest that you even consider this video format but since it is being used more and more it does deserve mention here, as does the conventional form.

A Video Storyboard

A video storyboard is just that—rolling videotape on your show and then looking at that tape to make the necessary alterations before shooting the show itself. It is an expensive process. You need your talent or standins for the taping or it makes no sense to do it, since without them the framing and action will not be included. Also, to make it accurate, you need to roll tape at the exact location, under the exact lighting conditions, with the sound and the music. You are in essence taping a complete scene. You may even want to edit this raw footage to time so that your video storyboard will present an accurate picture of what you want the video to look like.

As you can see, a video storyboard can become complicated and time consuming, not to mention expensive. Even so, some producers of big-budget films find them economical to do. It saves them millions of dollars on the other side. For you, however, it is probably not worth the time and effort or expense. You would be better served by the standard storyboarding technique.

Do not throw out the video storyboarding idea entirely, however. Rather, we would suggest that you adapt the concept to taping portions of your show. In other words, you want to take a camera to an exact location to tape a scene, with or without talent, just to ensure that the lighting will be right, for example.

A perfect example is a wedding. You do have an opportunity to create a minivideo storyboard during the rehearsals for the wedding. You can check your lighting, your framing, and your audio requirements just by rolling tape on this event. Careful review of this trial run or video storyboard will make it possible for you to deal with problems and make vital decisions prior to the taping day. And this extra footage could even be a bonus to the final version of the video, adding another aspect, the rehearsal.

The Complete Storyboard

Once you have given storyboarding a try, you may find that your storyboarding skills are so good that you can use the storyboards as your shooting guide instead of the format or script. In its most complete form, the storyboard, like the script, allows you to develop the idea to a finished visual with all the visual moves and transitions noted and all the audio noted as well.

Let's try this out on our wedding video. We begin with a wide shot of the church to establish the location, Figure 8–7.

From there, we go to a medium closeup of the bride and her attendants waiting in the church foyer, Figure 8–8.

And from there to a medium wide of the groom entering to the altar with his best man, Figure 8–9.

Back to a wide shot as the bride comes down the aisle, Figure 8–10.

And so on. In these storyboards, we have included all elements we need and, if we choose, we can actually shoot the video from them. One possible problem is dialogue. If you do have written dialogue, there is very little room to write it on the storyboard form. You can cope with that problem with a dialogue-only script keyed to the panel numbers on the storyboard or with a complete script given to only those crew members and performers that need to know what dialogue is being said. For the most part, this is only the producer, director, audio director and performers. Everyone else can work from the storyboard.

OPEN WIDE CHURCH.
MUSIC: WE'VE ONLY
JUST BEGUN.

Figure 8–7 Wedding video, storyboard panel #1. WS exterior of the church.

CUT TO MW CHURCH
FOYER.
SOUND: MUTED TALK-
ING, LAUGHING.

Figure 8–8 Wedding video, storyboard panel #2. MCU bride and her attendants in church foyer.

CUT TO GROOM AND
BEST MAN ENTERING
TO ALTAR.
MUSIC: FADE OUT.

Figure 8–9 Wedding video, storyboard panel #3. MW groom and best man entering to church altar.

CUT TO REVERSE SHOT.
BRIDE ENTERS AND
GOES TOWARD ALTAR.
MUSIC: WEDDING MARCH

Figure 8–10 Wedding video, storyboard panel #4. WS bride and her father coming down the church aisle.

CONCLUSION

Storyboards are visuals, using hand drawings of scenes or specific shots. Their purpose is to make certain that everyone on the shoot sees the show the same way. They are the director's way of conveying to the crew and talent the exact framing and action he wants recorded. You may create the conventional storyboard on paper or you can opt for a video storyboard on videotape if you have the time and money to create one. A complete storyboarding of a show can become the shooting script, replacing both the formalized script and the format.

Storyboards can be fun and if you are greatly intimidated by writing a format or script, they can be a nice visual replacement that still allows you to plan effectively. You can even start with the storyboard and then write a script. That is okay, too. The idea is to be aware of all the elements of planning and then pick those that work for your situation.

Coming up next: *The Final Pieces* . . . those last-minute details to be attended to before you can roll tape.

TO DO

1. Draw your own Storyboards using the form.
2. Roll tape on one of your locations, creating a video storyboard. You do not need to bring in the performers if you do not wish. You can roll tape just to get the shot framing and to test the lighting.
3. You have completed your Storyboard. Check this off your Master Checklist.

CHAPTER NINE

THE FINAL PIECES

Materials needed for this chapter:

Completed

FORMAT or SCRIPT or STORYBOARD

Still Working On

MASTER CHECKLIST

Blank Forms

LOCATION LIST
CAST LIST
PROP LIST

Now the real fun begins! Now you can begin putting the pieces together to make the video. Specifically, you can now begin fine-tuning the casting, selecting your locations, making prop lists, and determining whether you need a special set for the taping. You already have made some preliminary decisions about all these in your outline but now is the time to lock them in.

Some of these specifics may be dictated by the subject itself. For example, the main performers of our wedding video are the bride and groom. We have no control over that. And the main locations where we will tape are dictated for us also, chosen by the bride and groom. And, of course, we will not need a set for the wedding video, and the props, wardrobe, and makeup will all be provided by the occasion itself. In many of your videos, you may find some of the production decisions made for you by the occasion. In that case, the subject of the video will dictate rather than the producer or director. However, we would hope that you will expand your videos beyond these subject-oriented ones to purely creative ones over which you have control of all aspects of the production process. Specifically, we hope you will try your hand at producing one of the scripts at the back of this book or write a creative script of your own.

Traditionally, all the pieces that will make up the video like props, sets, locations, performers, makeup, hair, and wardrobe are locked in after the video is in its final form, whether that final form is outline, format, script or storyboard. So while you may have made some notes along the way and have a good idea of what you want to do in specific areas, now is the time to make your final selections.

WHERE CAN YOU TAPE?

Always approach the video with the idea that you can tape just about anywhere you want. There are few exceptions where you cannot. Some places do require you to get special permission, but these are few and far between and even then, special permission may mean nothing more than telling someone in authority at the location when you will be taping. Generally, anyone with a camcorder can go anywhere and tape anything without special clearance. You can, for example, walk out onto the Golden Gate Bridge or go to the top of the Eiffel Tower and roll tape, while the professional television producer must get special permits, hire police officers to control traffic, and pay a use fee.

Since you do have virtually unlimited access to really great locations, you have no excuse for having a visually boring video. Even in the most structured show, there is still room for some creativity. You can add locations for effect or to create a mood or just use them to make the beginning and end visually exciting. The number of locations you use is limited only by access to the location and, of course, time and money.

Access will be the biggest limiting factor. If you are in Iowa, the selection of the Eiffel Tower as a location is not realistic. Even if you have the time to do it, the money budgeted will probably not permit it. Or you may have the money but no time. Then again, you could have the time *and* money and be on your way to Paris to tape. Don't let access, time, or money keep you from at least considering certain locations. Better to weigh what a location will add to the video against access, time, and money than to just reject it out of hand. Don't be afraid to think beyond the obvious locations.

Our wedding video, for example, has certain locations dictated by the occasion:

Locations

Church
Reception hall

We have no control over these locations, but we have added three other locations that we believe will make the show more exciting. At one of these, we will be taping an interview, at another Mary buying her dress, and at the other we will be taping the opening of the show. The three locations we have added are:

Locations

Bride's home
Park
Department store

You should try to keep your locations varied to maintain visual variety in the show. Cutting from an interior of the bride's house to an interior of the groom's house would not be very visually exciting. Nor would it be exciting if these were the only two locations for the entire show. Variety will keep the show moving and help you create the visual excitement you need to hold the audience. Use cutaways and wide shots to help maintain visual variety.

If you have a location where lighting, sound, and action can be more easily controlled, this would be your studio location. This may be inside a warehouse designed specifically for television production or it may be a place like your living room where you plan to tape audio or graphics. A studio location needs only to be big enough for what is to be taped.

As you know, locations are designated as either exterior or interior. It is always a good idea to tape exterior shots of all your locations, whether they are called for in the script or not. You can then use these exterior shots to establish the location or as transition to the location. It helps the viewer to follow along and become involved in the action if they see where they are. You could tape the establishing exterior shots at any time prior to the taping date unless, of course, you need your performers in the shot. Then you can either get them to wear what they will be wearing that day and preshoot or wait until the day of taping to do the exterior.

For our wedding video, we can pretape some exteriors because our plan does not include either the bride or the groom or because it does not require that they wear their wedding attire. Pretaping these exteriors will allow us to concentrate more on the action on the day of taping. We can even pretape some of our interiors for the wedding, but we will get into this in more depth when we get to the shot list. For now, our objective is to decide our exact locations.

Once we have determined where we would like to shoot, we may want to pass these locations by others involved in the show or we may not. You are the producer and therefore, the ultimate decision should be yours. However, as a courtesy, particularly if it is a subject-controlled video such as the wedding, you may want to allow your performers some input. In our case, we may want to consult with the bride and groom regarding the two locations we have added since we are making the tape for them. While it is our show, it is still their wedding and courtesy dictates that we allow them to participate in how the show will look or at least to participate in some of the basic decisions.

How the show ultimately looks will, of course, be a product of not only the pieces that make it but your style in shooting and editing it. Ultimately, it is your show.

Make a list of locations, specifying interiors and exteriors, using the location list form, shown in Figure 9–1.

Our wedding location list would look like this:

Field Locations	*Exteriors*	*Interiors*
PARK	X	
MARY'S HOUSE	X	X
CHURCH	X	X
RECEPTION HALL	X	X
DEPARTMENT STORE	X	X
Studio Location		
DIRECTOR'S HOUSE		X

WHO ARE YOUR PERFORMERS?

Before you can begin taping, you must complete the casting. Every role, whether it is a major role or just an extra, must be cast. Everyone who is going to appear in the video must have a function, a purpose for being there, even in an unstructured, impromptu, live situation like the wedding. Everyone should know how they fit into the video whether they have one of the major roles or are just one of the extras. Knowing where they fit in will give even the most insignificant performer a sense of importance, thus giving them the incentive to be at their best when tape is rolling.

Your cast list will list everyone who will appear in the show including:

Principals. The major performers with the lion's share of the dialogue and action;

Secondary roles. Have speaking parts but smaller than those of the principals;

Announcer or voice over. Someone who narrates the action; may or may not appear on camera;

Extras. People serving as background visual and noise only. If you prefer, you can call them *crowd* or *audience;*

Groups. Groups of people that are doing something special for the video i.e. cheerleaders, a dance team;

Music. A band. A singing group. A guitar player. Anyone who is providing live music to the video.

Show Title _____

Date _____

Locations:

Interior   Exterior

LOCATION LIST

FIELD LOCATIONS   EXTERIORS   INTERIORS

_____   _____   _____

_____   _____   _____

_____   _____   _____

_____   _____   _____

_____   _____   _____

_____   _____   _____

STUDIO LOCATION

_____   _____   _____

Figure 9–1   Location List form.

Use the cast listed on your outline as a starting place to complete the cast list form shown in Figure 9–2. To this add any additional speaking roles. Then add extras, groups and musical accompaniment you may have worked into the script or that will be at the taping site as part of the occasion. Our wedding cast list would read:

BRIDE   MOTHER OF THE BRIDE
GROOM   MOTHER OF THE GROOM
PREACHER   FATHER OF THE BRIDE
MARY'S GIRLFRIEND   FATHER OF THE GROOM
MARY'S LITTLE BROTHER   KATHY, SINGER
JOHN'S BEST MAN   THE WEDDING PARTY

Show Title _____

Date _____

CAST LIST

ROLE _____   TALENT _____

ROLE _____   TALENT _____

ROLE _____   TALENT _____

ROLE _____   TALENT _____

ROLE _____   TALENT _____

ROLE _____   TALENT _____

ROLE _____   TALENT _____

ROLE _____   TALENT _____

ROLE _____   TALENT _____

ROLE _____   TALENT _____

ROLE _____   TALENT _____

ROLE _____   TALENT _____

EXTRAS _____

_____

_____

_____

_____

_____

Figure 9–2 Cast List form.

Once we have determined all the roles, we would begin casting. If you have natural stars, as our wedding does, this will be simple. If you do not, you may want to audition several people for roles. That is, you may want to ask them to read for you.

Conduct an audition by providing every possible cast member with a copy of the script, giving them several hours or a day to read it and become familiar with a part they might like to try out for. Then schedule a meeting for them to read for one of the roles. If you feel they can handle the role, then cast them. If not, you will want to cast someone else.

Diplomacy says that if you are going to hold auditions and you are dealing with friends that you had better plan an open audition where anyone can read for any role. Then the competition will not be so fierce and you will not be forced into casting someone who is wrong for the role purely out of friendship. If you keep the audition a general one, then you can cast the best people for the major roles and the less qualified ones in secondary roles with everyone getting a part. This approach will, hopefully, avoid hard feelings among friends.

There is a lot of ego involved in being on television. Be careful if you are in control of casting or you could end up with some bruised egos and lose some friends or colleagues along the way.

WILL YOU NEED MAKEUP, HAIR, OR WARDROBE?

In broadcast television, makeup, hair, and wardrobe are handled by crew members, not by the performers. In this way, the style of the show can better be maintained from one performer to another and from one scene to another. Also, it frees the performers from having to worry about their appearance, giving them more time to work on their role in the show.

Also, in broadcast television, everyone wears makeup, both men and women. That is why everyone looks so terrific. The makeup smooths out all the flaws. Hairdos may also be handled by the makeup artist or you may have a separate crew member that does just hair. Wardrobe can also be absolutely critical to specific roles. Take the pastel duo on *Miami Vice* or the high fashion chic of the ladies of *Dynasty*.

Makeup and wardrobe are an outgrowth of the show itself, its intent, its focus, its subject, its era. You may need extensive makeup and wardrobe or you may need none at all. The natural makeup of your female performers and the lack of makeup on your male performers may be just fine. Certainly, if you are doing a video like the wedding, you will not want to put makeup on everyone. It will be too time consuming for one thing and for another, the men probably will not allow you to put it on them.

On the other hand, if you are taping one of our scripts in the back of this book, do put makeup on everyone. It will make the show more professional looking.

In either case, it is always a good idea to have a powder puff and loose powder around during the taping to get rid of those shiny noses.

Makeup on television is generally very heavy, heavier than normal. What may look outrageous to the naked eye hardly even shows up on TV. If you are going to get into extensive use of makeup, it is advisable to learn more about how to do television makeup. In any case, if you are planning extensive makeup, be sure to test different looks before the cameras to make sure it works on tape the way you want it to. Also, since makeup and hair style go hand-in-hand, think hair styling also whenever you consider makeup.

If you want to create your own basic makeup and hair kit for the shoot, include in it the following:

PANCAKE MAKEUP (dark beige color)
LOOSE POWDER (natural color)
POWDER PUFF
POWDER BRUSH
DARK EYE LINER
DARK EYE SHADOW
RED ROUGE
LIPSTICKS
BRUSH
COMB
HAIR DRYER
HAIR CURLING IRON
GLITTER
REGULAR HAIR SPRAY
COLORED HAIR SPRAY
HAIR PINS
HAIR CLIPS

We have not included any special lists or forms for your makeup, hair, and wardrobe for the shoot. We leave this up to your own individual needs.

WILL YOU NEED ANY SPECIAL PROPS?

Props are important. What would Bogart have done without the Maltese falcon, the actual black bird? Or better yet, what would he have done without a cigarette hanging out of the side of his mouth? What would Colombo have done without his cigar? Or Kojak without his Tootsie Pop®? Where would E.T. have been without his Reese's Pieces®? All of these, the bird, the cigarette with matches, the cigar, the Tootsie Pop and the Reese's Pieces are Props, all necessary to create a desired effect or to further the action of the show.

Do you have props that are necessary to your video? If these are props that will be supplied by someone else, like the wedding ring, then you need not put them on your list of things to get. Be certain, however, that the other person who is supposedly supplying the prop is reliable. Otherwise, you could be ready to tape with everyone standing by and not be able to start the action for the lack of a prop. If you are unsure at all, you may want to provide a duplicate, just in case.

There is no need to list props that are not essential to the action or that will be at the location as part of the normal situation. If you want a book to be on a table and there is always one there, then you do not need to supply a book. However, if you want a particular book to be on the table, then it is wise to bring that book with you to the shoot. Just to be sure.

Make a list of the props you need for your video using the prop list form in Figure 9–3.

For our wedding video we have no prop list. There is nothing we will need that will not be made available to us by someone else as a part of the occasion.

Show Title _____

Date _____

PROP LIST

ITEM   NEEDED AT LOCATION

_____   _____

_____   _____

_____   _____

_____   _____

_____   _____

_____   _____

_____   _____

Figure 9–3 Prop List form.

DO YOU NEED A SPECIAL SET?

Generally, the answer to this question is *no*. But since this is television, there are always exceptions.

When you are talking set, you are talking about acquiring or constructing something, say a table or a wall or even an entire house. Most broadcast shows tape in a studio location and build virtually everything. If the show takes place in a house, they build the interior of the house that will be seen by the television audience. They then tape an exterior of an actual house, cut this to their studio built interior and you think that they are on location at the actual house.

Such a set is an enormous undertaking and costs in the five or even seven figures. Shown in Figure 9–4 are the plans for a set designed for a broadcast game show. It cost over $90,000 to build this set. You will not want to make such a huge commitment to a set, nor will you be able to afford to do so. Besides, you are able to tape at any location, unlike the professional broadcast show. Thus, your need for a formalized set will probably be minimal.

If you do feel the need to acquire or construct some piece of the set, borrow or beg it if you can. If you cannot, then draw up the plans, set a budget, a delivery deadline, and build it. This is not a complicated process but it can be costly if you are not careful. Even if you are only paying for the cost of the materials, it may be beyond the amount you want to spend.

If the video will not work without the set piece, then the decision is easy. You either make it or your forget making the video. If it is absolutely essential, then it is cheap at any price, if you are irrevocably committed to making the video. A good designer can help you determine ways to cut the cost of construction.

CONCLUSION

We have chosen our locations, cast our performers, made decisions about makeup, hair, and wardrobe, selected our props, and constructed any set pieces. We are ready to make the final check before we actually begin rolling tape.

Coming up next: *The Taping Plan* . . . the rechecking of all elements, the creation of a Shot List and the creation of the Show Bible.

TO DO

1. Complete your Cast List.
2. Note any makeup or wardrobe requirements.
3. Complete your Prop List.
4. Complete your Location List.
5. Complete set plans, if any.
6. Check Cast List, makeup, wardrobe, Prop List, Location List and set design off your Master Checklist.

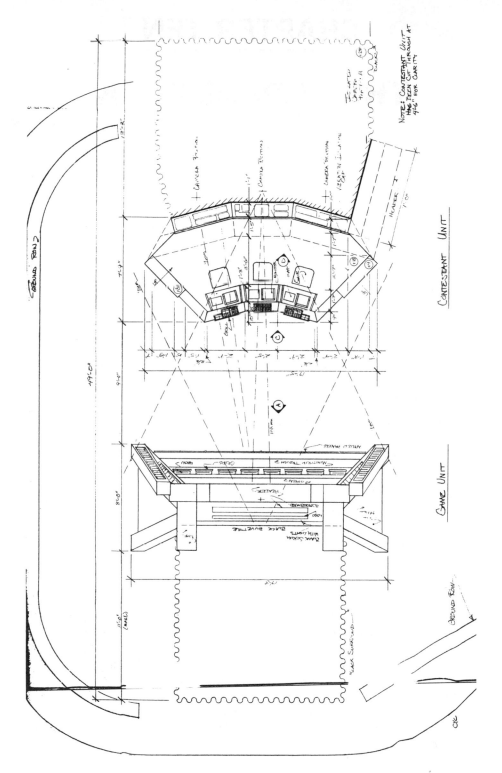

Figure 9-4 The set plans for a broadcast TV show. It cost $90,000 to construct this simple set for a game show.

CHAPTER TEN

THE TAPING PLAN

Materials needed for this chapter:

Completed

FORMAT, SCRIPT OR STORYBOARD
EQUIPMENT LIST
CREW LIST

Still Working On

MASTER CHECKLIST

Blank Forms

SHOT LIST

The taping plan is a combination of details that will make the taping sessions run smoothly and that will produce the best video possible. It includes preparing a shot list and the lighting plot, planning camera positions, making the final taping schedule, rechecking the equipment and crew lists and, if you have the time for it, setting up rehearsals for equipment and talent. This is the final planning stage prior to rolling tape. It is an exciting time, seeing the pieces come together, anticipating the finished video.

While it is possible to produce a tape without a format, even without a script, it is not possible to produce one without completing some form of the planning steps outlined in this chapter.

You *must* make a list of all your shots to make certain you record them all.

You *must* set your taping dates.

You *must* decide how many cameras you will have and where and how they will be positioned.

You *must* designate your audio feed, whether it is from the camera microphone, a remote microphone, or added in postproduction.

You *must* assess your lighting needs, arranging for additional or adjusting existing.

Finally, you *must* create the show Bible as your reference source on the shoot.

You probably will even rehearse, even if it is only testing a camera pan or asking your talent to begin a scene again. Follow the guidelines set down in this chapter and make certain you take full advantage of this critical planning stage of your video.

WHAT IS A SHOT LIST?

In television and the movies, shows are rarely taped in sequence. They are more likely to be taped out of sequence and then edited together in the right order, as shown in Figure 10–1. Time, convenience, and the availability of talent, crew, equipment, and location usually dictate when a particular scene will be taped, not the scene's position in the finished show. You could tape the end first and the beginning last. It does not really matter if you have an edit planned. It is smart to make use of your equipment and crew and tape as efficiently as possible by breaking the show down into specific shots and then grouping these shots according to location. Simply plan the shoot according to location, not according to shot order.

The list of shots according to location is called a *shot list*. It is simply a planned order for taping the show. With a shot list, you can tape every shot at a given location in one setup rather than running from one location to another to do the next shot in script order. A shot list can mean you go to a location only once, while if you taped in script order, you could be at the same location five or even six different times, every time having to set up the equipment all over again. This gets time consuming and expensive fast.

The shot list also makes it easier for you to make certain that all shots have been taped before you wrap the shoot for good. All you have to do is check shots off the list as they are shot. When all shots are checked off, you can feel confident about wrapping at that location and moving on to the next.

You can prepare a shot list from whatever you are using as the final form of the video—the script, the format or the storyboard. But do make a shot list. It will save you the time used in repeated setups at a location and the trauma of discovering later that you have forgotten to record a shot.

WALKING ON BEACH FOR CLOSE

EDIT TO

CHURCH EXTERIOR

CHURCH INTERIOR

Figure 10–1 Taping out of sequence is the rule rather than the exception in making TV shows.

We have included here a simple form, shown in Figure 10–2, that will help you prepare this important list. Take a look at it while we go through the specifics of how to make a shot list.

Numbering the Shots

If you have been following the plan outlined in this book, preparing a shot list is a relatively simple process. If you prepared a format or a script, you have already numbered your shots. All you need to do is make a list of these using the shot list form. For example, on the format for the wedding, park location, scene 2, prepared in the preceding chapter, we have numbered our shots as follows:

Shot #2–1 WS JOHN AND MARY WALKING IN PARK
#2–2 2S AS JOHN POPS THE QUESTION
#2–3 CU MARY'S ANSWER
#2–4 CU 2S AS THEY KISS
#2–5 WS PARK

Note that all shots to be taped in this scene begin with the scene number, #2. The main shot is numbered #2–1 and all others are numbered in consecutive order. By using this numbering system, we can easily see for what scene the shots were recorded. This will help us in the edit when we are making decisions about which shots to use.

If you have not designated any shot numbers on your format, script, or storyboard, then the first thing you want to do is to number them. Start at the beginning and number shots consecutively. If you were numbering the shots in a script for the first time, it might look like this:

#1   CUT TO WIDE SHOT OF THE EXTERIOR OF MARY'S HOUSE.

Number this shot #1 as shown.

#2   DISSOLVE TO INTERIOR MARY'S LIVING ROOM. FIRST PERSON PERSPECTIVE AS CAMERA WALKS ACROSS THE ROOM AND UP THE STAIRS.

<u>MARY</u>

(talking to Joan about the wedding)

ZOOM IN ON THE CLOSED DOOR TO MARY'S ROOM.

This is a new shot inside the house and thus numbered shot #2 as shown. You would not give the zoom another shot number. It is a continuing part of shot #2. The camera continues to move and therefore, even though the camera instructions may be broken up by dialogue, it is still all one shot.

Show Title _____

Date _____

SHOT LIST

LOCATION _____

Note: I = Interior   E = Exterior

Framing: CU, MS, WS, ECU, MCU, etc.

I or E	SHOT #	FRAMING	DESCRIPTION
_____	_____	_____	_____
_____	_____	_____	_____
_____	_____	_____	_____
_____	_____	_____	_____
_____	_____	_____	_____
_____	_____	_____	_____

CALL TIME _____

　　　　DAY _____

CALLED CREW (if all, just say "ALL")

_____

_____

CALLED CAST (if all, just say "ALL")

_____

_____

_____

_____

_____

Figure 10–2   Shot List form.

The rule to follow when numbering shots is: *Change camera angle or stop tape, start new shot number*. Number every shot on your script.

You may notice that there is a slightly different way of numbering shots when you are doing it from a script. You are numbering in consecutive order on a script, while on the format you are numbering by scenes, every scene having multiple shots, such as 2–1, 2–2, 2–3, and so on. Ultimately, there is no difference since both ways make it possible to create a shot list from which you can plan the taping. The one advantage the format numbering system has is that all its shot numbers relate back to a specific scene. The script numbering system, on the other hand, just numbers the shots in order. If you prefer to number scenes so you can use the scene number as a reference point in the edit, you can number the script by scene, and then break the scene into shots as you did in the format. It will just take a little more thought when you are doing it.

Putting the Shots in Location Order

Now that you have the shots numbered, you should break them up into groups, each group of shots representing a location. This will give you a list of the shots that you must tape at each location. You can then tape all the shots at any given location all at the same time, saving time and money.

For example, in our wedding video, all the shots at Mary's house can be taped at the same time with the exception of the wedding day events. All the church shots can be taped at the same time. All the park shots can be taped at the same time. If you must return for a special shot, as we must to Mary's house on the wedding day, at least all other taping at that location will be complete and you can focus on the special shot.

For example, let's say we have decided to number our script by scene as it was numbered in the format and have the following shot list for Scene #11, the Reception Hall location:

Shot List

LOCATION: RECEPTION HALL

I = Interior   E = Exterior
I   SHOT#11–1   MS   THE RECEPTION
I   #11–2   MS   JOHN AND MARY DANCING
I   #11–3   WS   MARY DANCING WITH RELATIVES
I   #11–4   CU   JOHN AND MARY CUTTING WEDDING CAKE
I   #11–5   CU   MOTHER OF THE BRIDE CONGRATULATING
                       HER DAUGHTER
I   #11–6   MS   JOHN AND HIS DAD

To this we would add the second scene planned at that location, Scene #12:

E   SHOT#12–1   WS   JOHN AND MARY LEAVING RECEPTION
E   #12–2   MS   JOHN AND MARY RUNNING BY
E   #12–3   MS   FRIENDS AND RELATIVES THROWING RICE

This would be our complete shot list for the reception location and we will tape all of it in one day. We do not need to tape even these shots in consecutive order. We could tape #12–2 first and then #11–5 or whatever. It doesn't matter. Again, you would decide based on convenience and availability of crew and performers.

By putting a description of the shot on the shot list, you can make this your master taping list and call for shots by number at every location.

Adding Shots Beyond What Is in the Script

We are not going to suggest that you tape additional shots just to be taping. This is self-destructive and will kill your edit with time-consuming decisions that should have been made prior to rolling tape. We *are* going to suggest that you tape some cover shots. You will already have a certain number of cutaways planned at each location but you may want to add more now to cover possible problems.

If you need a rule to follow, add no more than two cover shots for every specific scene. That should be more than enough.

Number these cover shots keyed to the number of the major shot for which they provide backup. For example, we may want to tape a CU of the wedding cake as a possible cutaway for Scene #11, the reception. We would number that shot #11–7, the scene number followed by the next consecutive number at that location for that scene. We would add it to the shot list as follows:

I   #11–7   CU WEDDING CAKE

We would know immediately from the numbering of the shot that it was taped as a possible cover for Scene #11. When we edit, we may or may not use it depending on whether the scene needs it. We could even decide to use it somewhere else in the show.

Using the Shot List

You will have a shot list for every location. Our wedding video has six different locations, so we would have six different shot lists. If you are not able to tape all scenes at one location on the same day for whatever reason, you will prepare two shot lists for that location, one for each taping day. For example, we are going to tape all our interviews and exteriors at Mary's house before the wedding day but we have to return to the house on the wedding day itself to tape the shots of Mary dressing. We therefore need two shot lists for Mary's house. One includes the shots we will tape before; the other lists those we will tape on the wedding day.

Once the shot list has been created, you will want to assign shots to specific cameras. Of course, if you have a one camera shoot, this will not be necessary. On a multiple camera shoot, however, you may want to designate which camera will cover what shot. This does not mean that you cannot have Camera #1 and Camera #2 tape the same shot, from different angles. The same shot from different angles is a great idea. It gives you a choice in the edit and choices are always preferable to being locked in.

To assign the shots, simply note in the margin on the left side, in front of the interior/exterior notation, the camera assigned to that shot. Note the camera by number, not the name of the camera operator. Then make copies of the shot list and give every camera operator one, keeping one for yourself.

As the shots are recorded, camera operators will check them off their list. You can use your copy of the shot list as the master copy, touching bases with the camera operators at various times during the shoot to see which shots they have checked off their lists, then checking those off yours. In this way, you will have a double-check system. The camera operators will be using their shot lists to make sure they have recorded all shots assigned to them on their copy of the shot list and you will be keeping a running update of all shots checked off all the camera operators' shot lists.

The shot list will give everyone a complete picture of exactly what you plan to accomplish at every location. They can all thus be up and ready to go when you roll tape. This list will also help you in directing the show. Since you will probably not be in direct contact with your camera operators when tape is actually rolling, this shot list will in essence direct them for you, telling them exactly what to shoot and even in what order if you elect to specify an order. By knowing what it is, the crew can anticipate the next shot and thus keep the taping session moving.

Remember, prepare one shot list for every location except when taping must be split up on different days due to the event.

WHEN WILL YOU TAPE?

What will be your taping schedule? Day? Time? Who is called for each taping date? Talent? Crew? Answering these questions is what scheduling is all about. Everyone must know what is expected of them and when it is expected in order to deliver their best performance, whether in front of or behind the camera lens. The shot list will tell them what is expected of them. The taping schedule will tell them when.

A taping schedule indicates both the time and day of the taping and what crew and talent must appear on that day and time. You will note that the shot list form shown in Figure 10–2 includes a place to indicate the call day and time, and a list of all those people called for that location. Fill in these blanks on the shot list for each of the locations. Then when you provide a copy of this shot list to everyone, you will also be notifying them of the taping schedule. If you are taping everything in one day, you should still indicate expected times of the taping at the various locations. It will help your crew and talent prepare themselves.

When you are scheduling the day and time of taping at a particular location, give yourself enough time to get the shots. If you expect to take two hours to tape at one location, give yourself two and half hours, just in case, especially if you are still somewhat of a novice at planning. Once you have mastered the art of creating great videos, you will be able to schedule more accurately.

Also when scheduling, plan everything around your major taping day, the wedding day, for example. We can begin production the day before or a week before or even a month before. We do not have to wait for that particular day to begin taping. By taping all the shots we can before or after the event, we will be freeing up more time on the appointed day, giving us more time to devote to the major portion of the show.

We would, as we have already noted, pretape all the scenes at Mary's house, the park, and the department store, none of which require Mary to be in her wedding gown. Also, none of these contain shots that can only be obtained on the actual wedding day. We could even pretape Mary dressing for the wedding if we chose to do so, assuming Mary was willing. However, since this scene has a certain energy generated by the day itself, we would not recommend this. While we could possibly get the same picture by pretaping, we would not be able to capture the excitement and energy that will be there only on the day itself.

How you schedule your shots can have an effect on how the show looks and whether it works. Keep this in mind when you are scheduling.

You should now be able to provide every member of your cast and crew with complete shot lists for every location that include the time and day of the taping session.

WILL YOU REHEARSE?

Now that you know when and where you will be taping, you can schedule a rehearsal of cast or crew prior to each taping if you feel the need. Rehearsals are important for establishing who is going to do what, deciding what the video is going to look like, and discovering any problems before committing anything to tape. But while they are good for all these reasons, rehearsals may be bad, particularly for amateur talent in a spontaneous situation like the wedding video. A rehearsal may cause them to tense up, making them more aware of the television camera than you want them to be. Generally, you do not want amateur talent in a real-life video to *act*. You want them to be themselves. You should, therefore, be wary of rehearsals, at least when dealing with nonprofessionals in this way. On the other hand, if you are taping a music video or a show of your own creation, rehearsal is a must since you want your talent, professional or otherwise, to do exactly what you have planned and scripted. A little rehearsal the day before the taping could save you hours in taping time and takes on the actual taping day.

Also, think about a rehearsal for your crew. This is particularly important for the novice crew members. While it may be detrimental to rehearse an amateur cast member, it is never detrimental to rehearse a crew member, amateur or the absolute best pro at the job. Even the professionals creating prime time broadcast television shows rehearse their crew down to the last camera move.

A crew rehearsal allows you to check not only the camera moves and your camera operator's ability to do them but the lighting, the audio, the location, and if you have the cast there, the props, the wardrobe, the makeup, and the cast members' performances. If you roll tape on a crew rehearsal, you can then schedule a crew meeting to view this tape, pointing out problems and noting where improvements can be made. This can give you access to valuable input from the crew while establishing a vital director–crew relationship.

We strongly suggest that you at least schedule a crew rehearsal, preferably at all locations, but at least at the major one. Follow this with the crew meeting and listen to everyone's input. It will make the show better and you a better producer and director.

WHERE WILL YOU PUT YOUR CAMERAS?

Whether you have one or ten cameras, you are going to have to decide where you want them positioned during the taping. Even a handheld, floating camera needs some guidance. Planning your camera positions is important to recording all the shots as you envisioned them. A poorly placed camera can make it impossible to capture a particular event, one perhaps lost forever if it cannot be repeated.

If, for example, in our wedding video we place the main camera in a place where it will be blocked when people stand, we could miss the bride's entrance or a portion of the ceremony itself.

Also, you want to place the cameras where they can give you the best variety of shots, thus working to their maximum ability.

Placement on the Location Sketch

To specifically show camera placement, first draw a sketch of the location. This sketch should be simple, as in Figure 10–3, noting only the items that affect camera or talent movement, like furniture, doors or windows. You need not include things like pictures that have no effect on the shot. Make the sketch as complete as it needs to be based on your taping plans.

Then simply indicate where each camera will be placed by drawing it on the sketch at the approximate location. You might want to use the camera symbols shown in Figure 10–4 on your sketch, just to be consistent. By using a standard symbol like these, you will also be able to immediately know what kind of camera position it is, handheld, tripod or dolly.

You or your assistant will use the completed camera placement sketch like the one in Figure 10–5 on the taping day to make sure that all the cameras have been set up in their proper places. To this sketch, we will add the positioning of microphones and lighting later in this chapter.

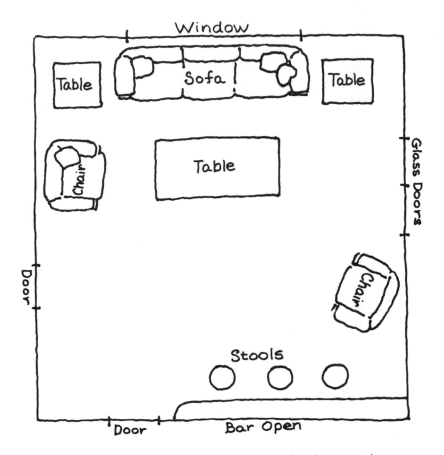

Figure 10-3   A location sketch should note only the basic furniture or other objects that will affect the movement of the cast or crew.

Choosing the Cameras

When deciding on cameras, ask yourself what kind of cameras you need. That is, how will they function? They may be handheld, mounted on a tripod, or moveable, mounted on a dolly. Refer to your equipment list where you have already made some preliminary decisions about this.

The Handheld Camera.   A handheld camera is carried by the operator and moves as he moves. If you have a handheld camera that has no permanent position but is a floater, put it on the sketch in its main area, if it has one. If it does not, put it at its first location. In either case, indicate it as a floater by putting an F following its camera number; for example, 2F inside its camera symbol on the sketch.

The Tripod Camera.   A camera is mounted on a tripod for stability. A

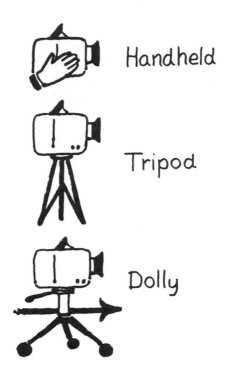

Handheld

Tripod

Dolly

Figure 10–4   Camera symbols you can use when you are placing your cameras on the location sketch.

tripod camera is thus not designed to be picked up and moved during a shot. It can make stationary camera moves such as tilt, pan, zoom, but it does not physically move its position during a taping sequence. That does not mean you cannot pick it up and reposition it, only that you do not want to do this while you are rolling tape on a particular shot.

Ideally you will want to position your tripod camera at one place at a location since it takes time to move. However, if you need to move it within a location, plan this moving time into the schedule. Note the movement on your sketch by putting it in its first position and then in its second position, labeling this second position by camera number, followed by the number 2, for example 1–2.

The Dolly Camera.   If you want the stability of a tripod but you need movement within the location, you should mount the tripod onto a cart with wheels, such as a wagon or skateboard or scooter. Anything that has a broad base and is strong enough to support the camera and its ancillary equipment can be a camera dolly. Once mounted this way, you can move through the location easily. The smoothness of your moves will be a factor of how smooth the floor or ground is, how skilled your camera operator is in operating a moving camera, and how skilled the grip or camera assistant is at smoothly moving the dolly from one position to an-

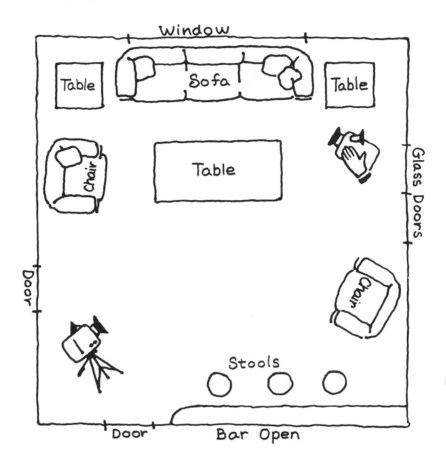

Figure 10–5 The location sketch with the cameras indicated in their positions.

other. Generally, a dolly camera will have a specific path that it will move along, rarely if ever leaving that path.

Indicate a dolly camera on the sketch by drawing it in at either end of the dolly path and then connecting these with arrows, as shown in Figure 10–6.

Numbering the Cameras

Begin your camera placement planning by numbering your cameras #1, #2, #3, and so on. This number will become the camera's identifying number for labeling, logs, or whatever. Also, you can call for shots by camera number instead of by operator name which is the way it is traditionally done by the professionals. Calling by camera number is easier and faster than calling out names. Besides, you may have two cameramen named John but you will never have two cameras numbered #1. Refer to your equipment list where you have already begun numbering your cameras.

Figure 10–6 How to indicate a dolly camera on the sketch.

Roll Tape at Location

It is best to plan camera placement by actually going to the location and looking at particular angles. This is the only sure way to make certain that the placement is a good one. If you can actually place a camera in the proposed location and roll some tape to get an exact picture of what angles you can capture at any one location, it is even better.

You might do this during a location survey trip or at a rehearsal. Take just one camera. Roll tape at different locations, recording at least a wide shot and a close up. As you begin rolling tape, say which location you are taping from. This will record on the tape with the picture. Later you can review the tape and make your decisions regarding camera placement. Better yet, take a color monitor along and review the shot right there and make your camera placement decisions on the spot.

WHAT AUDIO WILL YOU NEED?

You will have already made some basic decisions about your audio requirements when you made your equipment list. Consult this and adjust it to reflect your needs at the specific location. We will add to our location sketch, which already includes camera placement, any additional microphones and their placement.

The Instruments

There are, as you remember, five basic kinds of microphones available to us:

1. The camera microphone
2. The handheld microphone
3. The mikestand mounted microphone
4. The boom microphone
5. The Lavaliere microphone

The decision of which we need for the scene should be based on both the action taking place and our style in shooting it. For example, if we want no micro-

Boom   Handheld   Mikestand   Lavaliere

Figure 10–7 Microphone symbols you can use when you are placing microphones on the location sketch.

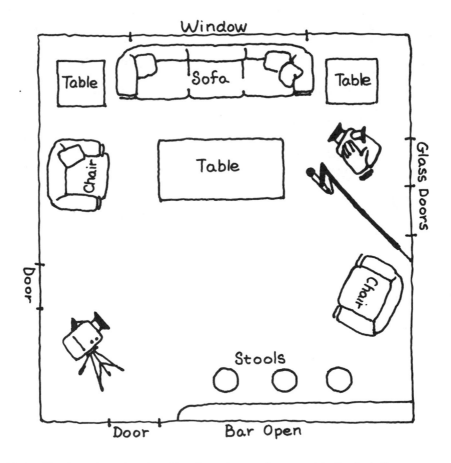

Figure 10–8 The location sketch with microphone placement indicated. Camera placement noted in Figure 10–5 remains.

phones to show, creating the illusion of reality, we would use the camera microphone if it was close enough to record decent sound, or we would use a boom microphone that could be held overhead, out of the shot. On the other hand, if we are taping an interview session, we might want to use "reporter style" and give our interviewer a handheld microphone that can be used by all, or we might attach Lavaliere microphones to the clothing of all those performers with speaking roles.

Choose the microphone based first on what you need to get the best sound recording, and second, just as important, on how the microphone can add to your style of shooting the scene.

Placement on the Location Sketch

Use the symbols in Figure 10–7 to place your microphones on the location sketch, as demonstrated in Figure 10–8. If you are using a camera microphone, there is no need to note this. We will assume that you are always recording sound on all your cameras unless you specify otherwise. There is certainly no reason not to record audio on all cameras. After all, you never know when you might need it.

WHAT IS A LIGHTING PLOT?

A *lighting plot* is simply a plan of where existing light is located and where you are going to place any additional lighting instruments. In the case of your video, it includes also any lights that you plan to mount on the cameras. While today's camcorders have been designed to adapt well to most lighting conditions, you will probably find that you need some additional lighting for certain locations.

A Brief Lighting Lesson

Without getting too much into the art of lighting television, there are basically two types of light:

1. Available light
2. Added light

Available light is what is already at the location. Sunlight, for example, or a lamp or an overhead light fixture or the light coming through a window. It is the light you have available to you without any effort on your part.

Added light is just that, any light you add to the scene. This can be another household lamp you bring in, professional video lighting instruments, or a shiny board.

These two types of lighting are positioned to provide certain kinds of light to the scene. Basically, they provide:

Key light
Back light
Fill light

A key light illuminates the scene on the side facing the camera; the back light outlines the scene or illuminates the background; and the fill light fills in any shadows that are unwanted. Key light is positioned at the front; back light behind; and fill light to the left and right, as shown in Figure 10–9.

Figure 10–9   Key light, back light, fill light: how they are used to light the subject.

This combination creates the basic "flat lit" scene where everything is lit and there are no shadows. If you want more dramatic lighting, you may allow a few shadows or you might want to put colored gels over some lights, creating a mood for the scene with blue, red, or yellow. Or you may want to include a light strobe pulsing different colors. You can do a lot with light if you just give it some thought.

Like camera placement, it would be of great benefit for you to roll tape at the locations to determine the lighting requirements. If the location is outside or inside with lots of windows, be sure to do this taping at the same time of day that you will actually be taping. This way you are more likely to get an accurate recording of what the available light will be.

Also, be sure to turn on all lights that will be on the day of taping. It would be helpful if you have a color monitor available at the location during this test taping so that you can immediately review the tape and test variations by adding to or adjusting the light.

The Instruments

What kind of lighting instruments will be available to you? Perhaps you have only what is normally available in the home or maybe you have access to professional lighting instruments. Since you are a novice at lighting, the best way to determine what effect any light is going to have on a scene is to put the light there, turn it on, and roll tape.

If you have access only to normal lighting like lamps or overhead lights, you can adjust this lighting by removing a lamp shade or adding one, by changing the bulb to a smaller or larger one, by adjusting it with a variable switch, or turning it off completely.

Just because the light is there does not mean you want or need it in your shot. If it does not add anything and certainly if it distracts from the show, *turn if off*.

During a daylight hour taping, a major source of light inside is windows. Many times this light is detrimental to the shot because outside light is brighter, making the inside picture you are trying to shoot darker, as seen in Figure 10–10. You can compensate for this in several ways. You can cover the window with drapes or blinds or you can put a thin, neutral colored film called neutral density filter (ND) over the window. It will screen the bright outside light, making it compatible with the inside light so it can add to the available light.

Drapes Opened

Drapes Closed

Figure 10–10   Window light may be too bright, causing your subject to disappear into darkness.

Outside you may want to make a shiny board for key or fill light. A shiny board may be a piece of cardboard covered with tin foil. This is held by someone to reflect the sun's rays onto the object you are taping. It provides a great deal of light and is extremely effective when used to provide fill light to a person's face, filling in the shadows caused by overhead sunlight.

Keep in mind that you are interested in the light on the person or object you are taping, not the camera location. Your camera may be 20 feet away from the taping site and in total darkness. It will not matter to what is recorded. Only the light that the camera is seeing in the picture you are recording is important. You do not, therefore, have to be concerned about lighting the entire location but only those areas where the camera will see action.

Placement on the Location Sketch

Once you have determined the best locations for all your lighting, take the location sketch (which already has on it camera and audio placement) and add to it the light placement. Whether it is an overhead or a lamp or a window, indicate on your sketch what you want from that light source during the taping. You can do this easily with a code system and eliminate the need for fancy drawings. Use the symbols shown in Figure 10–11 or make up your own.

Place your light symbols at the exact locations on the sketch. If you can determine the light output of each unit, all the better. If you have a light meter, use it. The lighting requirements of your camcorder will be in the manufacturer's instruction manual. Use it to help determine just how much light you will need to get a good picture.

Lighting is a subjective thing. It is a reflection of your style. It should be treated with respect and planned just as all the other elements are planned. If you

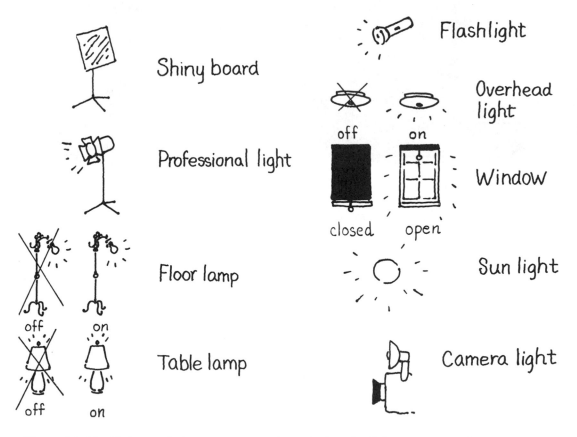

Figure 10–11 Light symbols you can use when placing the light sources on the location sketch.

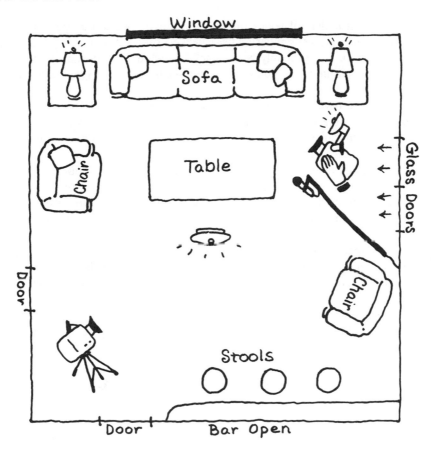

Figure 10–12 The location sketch with lighting positions indicated with their intensity, if possible.

just roll tape never looking at the lighting, you may have great shots, a great crew, a super story to tell it but you may not be able to see it because the lighting just does not work. The show is a function of all its pieces, all of which must be planned and executed properly. Lighting is one of those pieces, a very important one.

Your sketch now has camera and microphone placement as well as the lighting plot and looks like Figure 10–12. We will use this on the taping day to position these elements as planned.

WILL YOU HAVE A VIEWING ROOM?

A *viewing room* is a quiet place at the taping location where you can go to view recorded tape. You will find it of immense help in maintaining control of the shoot in terms of what is actually being recorded.

It is true that all your cameras record in color and that you can go and actually look through the viewfinder and see the recorded action playback. It is also true that this playback will probably be in black and white. If you want to see it as it is actually recorded, in color, you must have a separate color TV/monitor on which to view the tape.

If you have only one camera, you can set up this TV/monitor near your camera location or on a moveable table that you can move with you from location to location so you can view the tape at will. We don't recommend this moveable setup since it severely limits your camera mobility.

It is better to have a special viewing room where you can take the recorded tape to playback. That way, if you have a VCR in this room, you can leave your cameras on the floor still recording while you view tape. You also have the option of using the cameras as playback, bringing them to the viewing room. The choice is yours.

A special viewing room will also give you a quiet atmosphere away from the crowds and their influence so you can make those critical producer/director decisions. As seen in Figure 10–13, there are two ways to go in setting up your viewing room.

In Setup #1 there is just a color monitor with the camera used as the playback machine. With this setup, you will be plugging and unplugging every time you want to preview tape and you will be taking your cameras off the action while you are

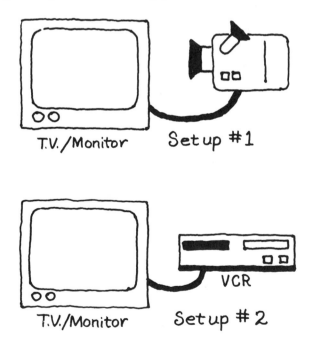

Figure 10–13 A viewing room may include just a color TV/monitor or it may include both the TV/monitor and a VCR for playback.

previewing. It does not take a lot of time to plug and unplug, but it does take some time, and that is time away from the taping itself.

In Setup #2, you never have to plug or unplug at all. The room is complete in itself with TV/monitor and VCR ready to go at all times. All you need do is insert the tape and push *play*. Your cameras never have to leave the floor but can remain on the action at all times. For obvious reasons, we strongly recommend this second setup.

Plan a viewing room. The small amount of time it takes you to plan its location and set it up will be well worth the time and worry it saves you on the other side.

WHO GETS WHAT?

The cast and crew do not need to go through the entire planning process with you but they do need to know what your final plans are relating to them and what their specific participation in the video will be. It will help them to understand their role as well as get them more involved in the production. Production information is provided to all on a need-to-know basis. The rule is: **If they need the information in order to perform their assigned duties, then give it to them. If not, do not burden them with paper that may possibly confuse them.** The things you might want to provide cast and crew include:

Format, Script, Storyboard.   You will not need to give them all these. We mention all three only because you may have elected to prepare one only, rather than following the complete formatting, scripting, and storyboarding process. Provide whichever of these is your *final* planning product. As a rule, only talent with speaking roles needs this information in addition to the director and the audio director, if you have one on the shoot. It can get costly if you start duplicating for everyone.

Crew List or Cast List.   If their name is on the list, they get a copy. This will confirm their roles in the taping and make them aware of the others involved and what their respective roles in the video will be.

Wardrobe/Makeup and Hair/Prop/Set.   If you have special requirements in wardrobe or makeup and hair for a particular cast member, you should make them aware of this. Others do not need to know. If you have a specific prop or set requirement, you should notify the crew responsible for this task. The cast members using the item need to know only that it will be there and how it will be used.

Sketch.   The sketch noting the camera and audio placement and the lighting plot should be given to the crew only. The talent does not need to know the technicalities. They will become aware of camera and microphone placement and lighting during the rehearsal or on the taping day.

Shot List with Taping Schedule.   This should be given to all crew and talent. It will ensure that everyone knows exactly what is expected of them on any given day, time or location.

Delegate the job of providing these forms to the appropriate people to an assistant. This assistant then can become the person in charge of making certain that everyone is in the right place at the right time and you can spend your time on the production side.

HOW DO YOU MAKE THE SHOW BIBLE?

We have now concluded all the planning prior to taping and it is now time to pull everything together into the three-ring binder that we asked you to provide back in Chapter 2. This will become the Bible for the show—your guide during the taping and the ultimate source for the answer to any questions. Anything not in the Bible

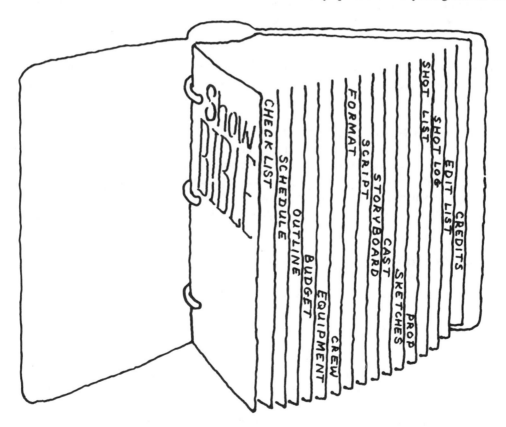

Figure 10–14   The Bible for the show is **all your plans** gathered together in one place, for ease in referring to them before, during, and after the shoot.

does not exist for your show. Your Bible should include tab locations, as shown in Figure 10–14, as follows:

MASTER CHECKLIST	STORYBOARDS
SCHEDULE (if you have a separate	CAST LIST
one from what is on the shot list)	LOCATION LIST AND
SHOW OUTLINE	SKETCHES
BUDGET WORKSHEET	PROP LIST
EQUIPMENT LIST	SHOT LIST
CREW LIST	SHOT LOG (blank forms)
SCENARIO AND FORMAT	EDIT LIST (blank forms)
SCRIPT	CREDITS

If you have elected to skip a step in the planning process and have nothing to put in a particular section, do not put a tab into the binder for that material. The shot log at this point will be blank, to be filled in during the actual taping. Also, the edit list is not yet prepared but you will need a place for it in the Bible because you will be completing it before you go to the edit session. This is also true for the credits.

Check over the Bible the night before the taping to make sure that everything is there. And, of course, take it with you to all taping sessions and to the edit.

CONCLUSION

Now that you have made your final decisions regarding camera and audio placement, lighting equipment and placement, and whether to include a viewing room at the taping site, you may have added to or taken away from your equipment and crew requirements. Now is the time to review your equipment list and your crew list to make sure they are accurate. As a matter of fact, it doesn't hurt to review all material in the Bible to make sure it is complete and up-to-date because we are about to **roll tape.**

Coming up next: *The Taping Day* . . . and we put all our plans to work at last.

TO DO

1. Complete a Shot List, with a separate piece of paper for every location. Assign shots to cameras.
2. Plan a day to run a test tape at the various locations to determine camera placement and audio and lighting requirements.
3. Complete a sketch of every location adding the camera placement, the audio requirements and the lighting plot.
4. Schedule the taping dates.
5. Prepare the Show Bible.
6. Check the Shot List, schedule, location sketches, lighting plot, camera positions and Bible off the Master Checklist.

CHAPTER ELEVEN

THE TAPING DAY

Materials needed for this chapter:

Completed

THE SHOW BIBLE

Still Working

MASTER CHECKLIST

Blank Forms

SHOT LOG

Finally, we have reached the day to roll tape on the video. We are out of preproduction, the pure paperwork portion, and into the actual production itself, the taping portion. Depending on the complexity of the video, this may be the only taping day or the first of many. Whichever, you are now going to use all your plans to make sure that what you record on tape is the best given the subject and circumstances behind the taping.

You can expect the absolute best if you have total control of the situation, including the subject, the locations, and the performers. If you have only partial control, such as in our wedding, then you can expect the best possible given the unknown events that may occur during the taping. Whichever your case, you can rest easy knowing that you are as ready as you can be, having spent time in preparing for the taping by making detailed plans. Now all you need to do is put those plans to work.

WHAT DO YOU DO ON TAPING DAY?

Before you can begin to roll tape, you must make certain that everyone and everything is in position and ready to go. This is a simple process that requires a minimal amount of time, but it is very crucial time spent in checking lists or executing plans already made. You will have everything you need in your Bible, the three-ring binder you have prepared. Be sure to take it with you on all taping days.

Begin your taping day by arriving at the appointed location at least half an hour before the call time for cast and crew. You are, after all, the director. It is your duty to set an example for everyone else. If you are sloppy about living up to the plans for the taping, your cast and crew will be sloppy as well. If you treat the production as a serious taping, your cast and crew will do the same. If you act like a professional, they will do their best to give you everything you ask of them. After all, they want to be professional, too.

The first thing you will do at the taping site is to use the Bible to make sure everything you need is there and ready before you roll tape. Make your check in the following order:

1. EQUIPMENT.   Using the equipment list, make sure that all equipment is at the taping site. If something is missing, send the go-fer for it if there is time. If there is not enough time then rethink the positioning of equipment and try to make do without it. Be sure that the blank videotape is at the site.

2. CREW.   Check off the crew as they arrive, noting next to their name their time of arrival. If anyone is slacking off, arriving late, impress on them the importance of keeping to the schedule. A camera operator arriving ten minutes late could mean ten minutes more before you can roll tape and if this is a spontaneous event, that might mean you will miss some action you wanted to record. You can make this crew check on your crew list or on the shot list. Give all camera operators their allotment of blank videotape.

3. LOCATION SKETCH.   As your crew arrives, you should put them immediately to work positioning the cameras, the microphones, and the lighting according to the location sketch you made earlier.

4. CAMERA SETUP.   Once the cameras are in their designated positions, the camera operators should white balance their cameras and make certain that they are running properly. The white balancing can be done with a piece of white paper, according to the instructions provided by the manufacturer of that camera. To make sure that the camera is functioning properly, roll some tape and take a look. In addition to testing the camera, you will also be checking lighting and microphones. View the camera test recording and make any needed adjustments to cameras, lighting, and microphones.

5. VIEWING ROOM.   Set up your viewing room if you are going to have one. Use the camera test recording to make sure the playback equipment is working properly. Use the cameras as the playback machine during the equipment test as a further test to make certain they are operating properly. For viewing during the day,

ideally you will have a VCR dedicated to the viewing room so you will not have to pull cameras from the floor.

6. CREW REHEARSAL.   While you are waiting for the cast to arrive, take the opportunity to rehearse your crew by going through several camera moves. Every camera operator should have the shot list you gave them previously. If they do not, give them another copy. Then call for the first shot on the list. Go from one camera to another, looking through the viewfinder to make sure that the camera operator has the right framing for that first shot. You can ask the operator to do the second, the third, or however many shots you would like to preview. The number of shots you wish to see should be related to your experience with that operator and their past experience in using that particular camera. Once you are assured that the operator knows how to use the camera and knows how to read your shot list, then you can trust him or her to give you the shots you want. If you have time, you can roll tape on this camera rehearsal and review the material with your crew in the viewing room critiquing their shooting style in relationship to what you see for the video.

7. PROPS.   Check to make sure any needed props are at the taping site and in position, if appropriate. Make sure all functional props work as they should. For example, does the cigarette lighter work? Will the radio play? Will the telephone ring?

8. CAST LIST.   As your cast arrives, check them off the cast list, noting their time of arrival. And, as you did with the crew, reprimand those cast members who do not arrive on time. The obvious exception to this cast check is if you are taping a spontaneous event, such as the wedding, then your cast will arrive whenever they want to and certainly you have no cause to criticize them. If you are providing makeup, hair, and wardrobe, send all cast members there as they arrive.

9. CAST REVIEW.   Make sure that all cast members have their shot list just like you did with the crew. Take a few minutes to go over the shots with the cast. If you cannot cover them all, at least review the first two or three. Even if this is a spontaneous video, you will want to do this shot review with the principal on-camera performers. It will help them know exactly what your taping style is for the event and how they fit into that style.

10. FULL REHEARSAL.   If you have the opportunity, it is a good idea to rehearse at least the first shot with all the elements in place, including cast, crew, camera, audio, and lighting. By recording this on tape, you can preview it in color on a TV or monitor to determine if you need to adjust one or all of these crucial elements. Hopefully, you will have time to rehearse every scene either all at once before first tape rolls or perhaps immediately before each is recorded. If you do not have time to rehearse them all, do the first one or two to at least make sure the overall lighting is sufficient and you are recording good quality audio.

11. FORMAT/SCRIPT/STORYBOARD.   Depending on your video and which of these you have prepared, you will want to have one or all of these available on the taping day for yourself and possibly some of the crew and cast. It will clarify any lingering questions about the style while being a quick reference for those who

need to know how the events relate to one another and what comes next. Decide who needs this and make sure they have it.

When you have accomplished all the above, you are ready to **roll tape.**

WHAT TO DO WHILE ROLLING TAPE?

Once you are rolling tape, you, as the director, need only the shot list, which you have already prepared. If you have a formalized script with dialogue, like one of the scripts at the back of this book, this shot list may be written directly onto that script so that in addition to making certain all shots are recorded you can make sure your performers deliver all their lines in the manner called for in the script. You will use the shot list as the source for all shots to be recorded, as a checklist to monitor the work of the camera operators.

The Shot Log

All camera operators will need not only a copy of the shot list but also several copies of the shot log form shown in Figure 11–1. The camera operator, working with her or his assistant will use the shot list as their guide for the day's taping. The camera operators know that all shots on the shot list assigned to them must be recorded before the director calls a wrap for the day. Shots will be checked off this list as they are recorded to make it easier to know when the list has been completed.

As they are recorded, all shots will be noted on the shot log indicating exactly where the shot is recorded on the tape by using the cassette number and the camera counter number. Also, the shot log will include a special comment on every shot, noting, for example, whether it is a good shot or not worth looking at again. Every camera will have its own separate shot log. Ideally, you will have an assistant for every camera operator to keep the log just for that camera. The logs will be more complete and thus easier to use if you have this extra person to assist, since it will be difficult for the camera operator to keep copious notes while trying to operate the camera properly.

However it is done, you *must* have a shot log for *every* camera. It is the only way to save endless hours of viewing tape in postproduction.

On the shot log, the assistant will write at the top of every page:

Show name: Title or proposed title of show.
Date: Taping day and date.
Producer: Name of producer. You?
Location: Name of location.
Page number: Page number in relationship to total number of pages, such as page 1 of 4.

SHOT LOG SHEET

Show Title _____

Date _____

Producer _____

Location _____

Page _____ of _____

Camera No. _____

Camera Operator _____

Camera Assistant _____

CASSETTE #	SHOT #	TAKE #	COUNTER NUMBER		APPROX TIME	NOTES
			IN	OUT		

Figure 11–1 Shot Log Form.

Camera number and operator's name: The camera's designated number and the camera operator's name.

Production or camera assistant: Name.

And then, for every shot:

Cassette number.   The number of the cassette being recorded on. All cameras should number their cassettes starting with number 1. The camera number will provide the differentiation between cameras.

Shot number.   Take this number from the shot list or if it is a newly created shot, give it the same number as the shot it follows and then a consecutive number. For example, shot #24 on the shot list might be followed by #24–1 and that shot followed by another extra shot numbered #24–2.

Take number.   The take number. If you shoot the same shot more than once, number each of the shots as takes. The first time you shoot it is Take 1; the second time Take 2; the third time Take 3; and so on.

The counter number.   Indicate both the start or in number and the stop or out number. Take this number from the counter number on the camera or VCR. When you play the tape back, these numbers will probably not be perfect, but at least they will get you close to the shot on that piece of tape. Remember to zero the counter before you begin taping on a new cassette. Do not zero the counter in the middle of a cassette or the numbers will no longer be of any value. The stop or out time will be the number that is on the counter when you actually stop tape on a shot or when you begin framing for the next shot.

Approximate shot time.   It would be helpful for the assistant to have a stop watch to time every shot, noting the time on the shot log. This time estimate would enable you to determine the approximate length of shots and thus the show itself when you are planning your edit. While this is nice to have, it is not essential. Therefore, do not burden the camera operator or the assistant if the logistics of doing it are just impossible.

Notes.   Finally, anything of interest in the shot should be noted. A note can be as simple as *Good Shot,* to something more specific like *Great Closeup of Mary.* This note will help enormously in planning the edit.

One of our shot logs from the wedding video might look like that shown in Figure 11–2.

Remember your show is only as good as the shot logs. It is the only thing that counts after you have wrapped the shoot. With it, you have a concise, descriptive list of all shots. Without it, you have a stack of cassettes with perhaps an idea of what is recorded on each of them but no idea where to find it. The log is the only thing that will allow you to find the best shots, using them to make a better video, all as painlessly and as quickly as possible.

Show Title _Wedding Video_

Date _11 – 4_

Producer _You_

Location _the Church_

SHOT LOG SHEET

Page _1_ of _5_

Camera No. _2_

Camera Operator _James_

Camera Assistant _Tootsie_

COUNTER NUMBER

CASSETTE #	SHOT #	TAKE #	IN	OUT	APPROX TIME	NOTES
2	15 – 2	1	0015	0065	:30	good stuff wide with Mary arriving
	15 – 3	1	0075	0085	:10	Exterior good CU Mary's face
	15 – 4	1	0095	0124	:20	Wedding party greeting Mary
	15 – 9	1	0134	0144	:30	Interior CU Mary laughing
	15 – 7	1	0154	0180	1:10	Mom with Mary good stuff
		2	0190	0267	2:00	more Mom and Mary Straightening veil, etc.
	15 – 4	1	0277	0312	:30	good Dad and Mary
	15 – 11	1	0322	0340	:20	CU Dad – good stuff

Figure 11-2 One page of wedding video shot logs from Camera #2.

The Taping Procedure

To make sure that all shots are recorded, are of acceptable quality, and are logged properly, get your camera operators to follow this simple procedure during the taping day:

Camera Operator's Checklist

Note: *You* refers to the camera operators when used on this list.

1. Give the cassette a number. Label it and the box it comes in. Tell your assistant the casette number and then insert the cassette into the camera.
2. Zero the counter on the camera.
3. Roll 30 seconds to 1 minute of tape to make sure the camera is recording properly.
4. Tell your assistant the number of the shot you will be taping as it is noted on the shot list.
5. Frame the shot in the camera and practice any moves you will be making.
6. Roll tape, recording at least 30 seconds *before* the needed shot begins.
7. Tell your assistant the counter number on the camera as the shot begins.
8. Tape the shot.
9. Stop tape.
10. Tell your assistant the counter number where the shot ended.
11. Give the assistant your impression of the shot, its good and bad points, so this can be included in the notes on the log.
12. Review the shot, all or part, to make sure you had no dropout or technical problems on the tape.
13. If you recorded a bad or marginal shot and can redo it, roll tape on Take 2, noting in and out times on the counter.
14. Once you have an acceptable shot, check the shot off the shot list.
15. Look for cutaways. If you see any that you feel will be useful, give them a shot number, note counter times, and roll tape on them. On the log describe exactly what the shot is, such as CU wedding cake.
16. Go on to the next shot.
17. Remember to tell your assistant the new cassette number if you change cassettes and always remember to zero the counter on the camera when you insert a new cassette.

The Retake

Just because a shot is recorded on tape does not mean that you have a shot that is acceptable or usable or even desirable. Whether a shot is any good or not does not make a difference if it is not possible to do a second take on the shot, like the *I Do* in the wedding ceremony. You have no choice but to live with what you have recorded, possibly using a cover shot to make it better, or if it is not absolutely essential to the video, you could throw the shot out entirely. On the other hand, if it is possible to do a retake on a bad shot, by all means do it.

It would be wise to make it a practice to preview recorded material throughout the taping day, adding additional takes if necessary. If you do not have the luxury of reviewing recorded material and there is some question in your mind about whether a shot is good or not and the opportunity to do a Take 2 is available, then by all means do a Take 2. It will not hurt anything and it may save you later.

Allow the camera operators to initiate some second takes. Don't be afraid to rely on their judgment but always remember that their assessment of the shot is based on the black and white playback they see in their viewfinder. The picture may look totally different in color. Therefore, it is always a good idea for the director to spot-check tape in the viewing room where color playback is available whenever possible.

If you, as the director, have the opportunity to spot-check the recorded material during the taping day, look for the need for additional takes by following this simple procedure:

1. Review the shot, if possible, in the viewing room on a color monitor. Check the lighting, the audio, the framing, and the action.
2. Reshoot if the shot is not acceptable. Number this take, such as Take 2 or Take 3. Always keep all the takes. Do not burn even the worst of them. You may need it as cover material.
3. If it is not possible to do a retake, note on the log the problems with the shot and explore the shot for cutaways that might cover the problems. Shoot these cutaways, logging them and noting that they are possible cover material for the questionable shot.

This procedure will help you avoid some problems, if not all of them. The key to producing a good video is flexibility. Be ready to change, to create new ways to get the shots you want and need.

Before You Wrap

Before you call a wrap for the day, make certain:

1. You have on tape every shot you had planned to record. All shot lists should be checked off.
2. Your shot logs are as complete as they can be. Review these and have the camera operator or the assistant add to them if needed while the taping is still fresh in their minds.
3. You have gathered up all your equipment and it is stored properly including extension cords and plugs.
4. Everyone, cast and crew, knows when and where to report for the next shoot day, if there is one.

If you have time to spot-check the tape in your viewing room, do so just to re-assure yourself. Also, if you have another shoot day planned, showing the recorded material to cast and crew can charge them up for the next day. Or if it is a spontaneous video, participants will enjoy seeing some of this original material. Do not get carried away viewing tape though. It will lose its impact rapidly since raw footage can get boring fast. Bits and pieces chosen here and there will be more than enough to convince everyone, including yourself, that you have great stuff recorded.

WHAT CAN YOU DO TO RECORD BETTER SHOTS?

Here are a few basic tips that might make your taping go a bit smoother and give you better shots at the same time.

Cutaways/Cover

As you already know, these are the shots that you need to cover missed shots, bad shots or to provide a transition from one shot to another. Or you may just see an interesting shot you would like to include. These are the kinds of shots you will learn, with experience, to tape to make your life easier in the edit and to make a better video. In time, you will discover that you will call for cutaways spontaneously during the taping, as you "see" them.

Generally, cutaways are taped after a scene. For example, after the wedding ceremony is over, we may ask our bride and groom to hold hands, showing off the wedding ring, so we can tape a closeup of it. Or we may ask them to kiss again so we can get a closeup of that action. Cutaways are generally taped after the fact like this because, having already shot the scene, you have the advantage of knowing what your potential video problems might be and you also know what special action happened during the taping that might lend itself to a nice cutaway. Learn to look for cutaways like those shown in Figure 11–3.

Figure 11–3 A cutaway is shot to emphasize an action or to provide variety to the wide shot (WS) or to cover possible problems with other material.

Extreme Closeups

An extreme closeup, as shown in Figure 11–4, is just exactly what it sounds like, a very tight shot of whatever. The classic example is an ECU of a person's lips. An ECU can be intimidating, sensual, informative, or emotional. It is whatever you make it in the context of the show. Someone in a CU talking to the camera who leans in to an ECU can be threatening you or telling you a secret. An ECU of the bride's finger as the groom puts the ring on can be emotional. Look for opportunities to use the ECU effectively.

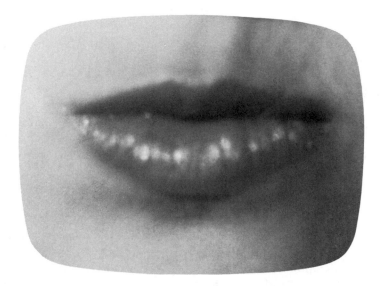

Figure 11–4   An extreme closeup (ECU) may be used to intimidate, especially if the talent leans into it, or it can be used to bring attention to something in the shot.

Wide Establishing Shots

While we all know that television is best seen in closeups due to the small viewing screen, wide establishing shots are also extremely important to any show. They give the viewer location and perspective and make it easier for them to follow the action. For example, if you are taping at the park and all we ever see is one of the cast members jogging down a path, we, as the audience, are not in the park at all. We are just jogging down a path. In order to put us in the park, you have to show us the park in a wide shot, as demonstrated in Figure 11–5.

Actually, you do not even have to tape the jogger at the park. If you show us a wide shot of the park and then show us the jogger, we will believe that he is in the park. He could actually be in your back yard. As long as we do not see anything in

Figure 11–5   A wide establishing shot tells the audience specifically where you are.

the background to give away the fact that we are in your back yard, we will believe he is in the park.

This effect is used all the time on broadcast television. You may see a wide establishing shot of a hospital, for example. Then the show cuts inside to a room. This room is actually in a studio, not in the hospital. But because you have seen the exterior shot of the hospital and because the interior shot looks like a hospital room, we believe we are inside that hospital.

You can use this on your shows, putting your stars anywhere you want by just editing in a wide establishing shot and following that with a reasonable interior of what the viewer believes the interior of that location looks like or following it with a closeup that doesn't allow the viewer to really see beyond the action. You could live in the most expensive mansion or fly an airplane or go to the moon when a wide shot puts you there.

In addition to creating an effect, wide establishing shots are important to the audience's understanding of the show. If we do not show them where the story takes place, they will not know. One park is like any other unless we establish with a wide shot that we are, say, in Central Park in New York City. That changes our perception of the scene entirely.

Interviews

Interviews add an element of spontaneity to a show, even if they are partially scripted. You can script the questions to be asked in the interview. You can even preinterview and then make it clear to the interviewee exactly which response you want delivered on camera. Or you can conduct a spontaneous interview on camera with no set time limit. We recommend you do the latter with an interviewer who is good on his feet and can pick up on responses and spontaneously formulate new questions. You can then take this lengthy recorded material and edit the interview to delete those portions that are uninteresting, redundant, or that just do not relate to the subject of the show.

Spontaneous interviews can be of particular interest and fun in a family video. It gives Uncle John a chance to be on camera, an opportunity he would as a rule shy away from entirely, but because you do not structure his responses, allowing him to just be himself, he feels comfortable before the camera and we are allowed to see a side of him we have perhaps not seen before. His half-hour interview can become one minute or less on the video, a minute that makes Uncle John look great. This type of interview makes the show more personal and more fun.

Anticipate the Moment

Too often when tape is rolling, we forget to think beyond the shot being recorded. Thus, great shots are lost. Do not be rigid about your plan for the show. Leave a little room for those unexpected moments and learn to anticipate, even predict them.

For example, let's say that while we were taping the wedding video, the bride and groom are dancing and suddenly the groom picks up the bride, twirling her around to the music. It is an emotional moment and totally unplanned by anyone. We certainly did not have this shot planned and not only that, we were getting set up for their departure on the honeymoon and were not rolling tape on them at all. But being ever ready for the unexpected at these unstructured events, on seeing what was happening, we immediately pulled a camera off its assigned place and panned around to capture this very special moment. We knew we would have ample time to set up again for the planned shot and, after all, the bride and groom are the stars of this show. It makes sense to record this special moment. Thus the moment was not missed.

Learn to anticipate moments like this by totally immersing yourself in what you are taping that day. If you are totally involved, you will be less likely to miss shots and more likely to capture the very essence of the event.

Point of View

The camera acts as the eyes of the viewers. They see only what the camera sees. You can thus create different effects with the camera by simply changing the point of view of the camera lens. There are three basic points of view that the camera can assume: the first person, the second person, or the third person, as noted previously in Chapter 2 and demonstrated in Figure 2–22. The point of view (POV) you select for the camera can have great impact on the audience response. If, for example, we put the viewer inside the car as the couple drives away from the reception, they now have a second-person POV. They are participating, enjoying the moment on a personal level with the bride and groom. If, on the other hand, we shot the scene with a third-person POV, the viewer would be part of the crowd, watching from a distance as the happy couple leave. And finally, if we show the audience the couple getting into the car and then cut to a shot from the driver's seat, seeing the bride but not the groom, hearing their conversation, seeing the crowd out the window as the car drives away, we have put the viewer into a first-person POV, actually seeing the scene through the eyes of the groom.

As you can see, you can make your viewer a participant or an observer by the way you use your camera. You can also create a style that is yours alone by the camera's POV. Shot with a single POV, the wedding is like any other. From various POVs, it is unique.

Putting Yourself in the Picture

In spontaneous videos, too many times you get caught up in creating the video and forget that you, too, are often a part of the event. You edit the show and guess what? You get to take credit for creating it but you never appear in it at all. Maybe you are the bride's uncle, aunt, cousin, brother, or sister. You know you should have been in the show but you did not have time to include yourself. Rubbish. Take time. There are many opportunities for you to get into the show. Plan them. It may mean that you do a walk on like Alfred Hitchcock did in all of his films, or maybe you will say a few words. But you will be there. Your signature will be on the show. You have a right to be there.

Consider doing an Alfred Hitchcock at the very least. There is nothing wrong with wanting to be on television or wanting to be a part of what you know is going to be a dynamite show, *your show*.

CONCLUSION

We have now recorded all the footage at every location using the location sketch to make sure all cameras, microphones, and lighting were placed properly.

We have checked our shot list to make certain all shots have been recorded. We have checked all cassettes to make certain they are numbered correctly.

We have made certain that we have recorded enough cover material in case we need it in the edit. We have recorded all the wide establishing shots that we need to establish location. We have put ourselves into the video.

We have gathered all the shot logs and have checked them against the shot list to make sure all shots are recorded on the shot log. We have scanned the notes on the shot log to make certain we understand them before we release the crew.

We have gathered all our equipment and stored it. We are ready to begin the process of turning the raw footage we now have into the show.

Coming up next: *The Edit* . . . pulling it all together to make a show.

TO DO

1. Follow the setup steps.
2. Use your Shot List and make your Shot Logs while taping.
3. Check the Shot Log off your Master Checklist.

CHAPTER TWELVE

THE EDIT

Materials needed for this chapter:

Completed

RECORDED TAPE from your shoot
SHOT LOG from your shoot

Still Working

MASTER CHECKLIST

Blank Forms

EDIT LIST

The day after a shoot there is a certain tendency to let up, to slack off, confident that you have the show on tape. This is the wrong time to celebrate. You do not have the show until it is ready to dub, until it is ready to be seen by your audience. If you let up after the shoot, you will lose your momentum, your involvement in the product, and possibly lose the show itself in the process. There are things you know at the end of the shoot that you will forget if you delay the final steps.

You are not finished when tape stops. Now comes the most important part of the production. Well, maybe not the *most* important, but certainly just as important as getting great shots on tape. Now is the time for postproduction.

Postproduction is a combination of deciding exactly how you want to cut together the shots you have recorded and editing these shots together.

Postproduction begins by double-checking your shot list.

DO YOU HAVE EVERYTHING ON TAPE?

Before you edit, you want to make sure first that you have recorded every shot you had planned to tape on the production days and then you want to make sure that you record any graphics, photos, titles, and so on, that you will need to put everything together.

First, take a look at the shot logs and the shot list.

Checking the Shot Log Against the Shot List

Now is the time to review all shots you planned and make certain that you have everything you will need. You do not want to go in to edit to discover that you missed some critical shot.

Begin by checking your shot list against the shot logs. You will have checked it before you wrapped the shoot but do it again just to make sure that all shots were recorded. Do this by following these simple steps:

The Final Shot Check

1. Check the shot list against the shot logs. Are all the shots on the shot list recorded on the shot log?
2. Questions regarding any shots? Contact your camera operators or camera assistants. Perhaps they can answer the questions.
3. Questions about the notes on the log? Contact the camera operator or the camera assistant.
4. Sit down and view the shot if you have any unanswered questions.

Reshoot

If everything did not go well on the production days or if you discover in your shot check that you have missed a critical shot, you may have to schedule a pickup day. That is, you will go back to the location with the necessary performers and recreate the situation so you can roll tape again.

Do not do this unless you absolutely have to have a missed shot. You will find that it is difficult to create the exact same situation twice. The ambience will be different. The people will not have the same energy they had. The lighting may be different. And so on. It is best to cover a missed shot with something else if possible. You can always change a scene to accommodate a cover shot or maybe you can delete that scene altogether.

Explore alternatives before you think about a reshoot. But, if it just will not work without that shot, then you have no alternative but to schedule a pickup day and hope for the best.

Postproduction Taping

It is not uncommon to tape some additional video or audio immediately prior to or in the edit session. As a matter of fact, it is standard practice. There is always something that is not yet recorded on tape that you will need in the edit.

Video. There may be some visuals that you did not record on the production days but saved for taping after. These may be special artwork, photos, titles, or credits. Record these now on a separate piece of tape. Call this your *graphics cassette*. Be sure and log all these shots as you record them just as you did on the taping days so you will know where to find them later.

You could eliminate the graphic cassette and record this video live during the edit session but if you decide to do this, be aware that it will take time to position things and roll tape. Thus, your edit session will be longer.

If possible, we recommend that you record these on the graphics cassette and save yourself this hassle in the edit. There will be plenty of other problems at that time to occupy you. Best to save your energies for them.

Audio. In addition to visuals, you may have certain audio you want on tape for the edit. How or if you can get this audio onto your video will depend entirely on whether you can do an audio-only edit with your equipment. An *audio-only* edit is the ability to add audio to a tape without erasing the video already recorded on that tape. Even if you are able to do this audio-only edit through your video equipment or a stereo hookup, be aware that to add the new audio you may be giving up the original audio recorded with the video. The only way you could keep both new and original audio would be to mix the two or to record them on different tracks on the tape. You will remember that you have two audio tracks available on the tape. Audio can be mixed or recorded on different tracks using an edit controller that has that capability. If you do not have access to this equipment, you could schedule some time at a postproduction suite. This second alternative may not be as expensive as you think, particularly if that is all you want to do in the post suite. If you are considering this option, call and ask for a cost quote before you throw the idea out entirely or settle for less than you really want. There are many video stores catering to the needs of the smaller video producer that offer these services.

Our aim here is not to try to solve all these audio problems for you. Rather, we want only to make you aware that it is possible to add new audio in the edit and, if your equipment will not do it for you, it is important enough to the show to explore other ways of doing it.

Additional audio may include music, perhaps a special song or some original music you have created yourself. Or you may want to add a voice-over announcer to narrate a portion of the video.

Again, you can add this new audio to your graphics cassette or you can do it live during the edit session. If you do the latter, be sure to block out additional time in the edit. It will take this extra time to get the announcer rehearsed and recorded or the music cued and ready to roll. One of the benefits of doing it live at the edit is that you can put the entire show together and then match the audio to this video, making it fit perfectly according to your plan.

HOW DO YOU GET READY FOR THE EDIT?

Before you go to the edit, whether it is in a professional edit studio or your own facility, gather up all the elements you are going to need in the edit. You may have one or all of the following:

1. Recorded video cassettes—the cassettes you recorded on your taping days
2. Graphics cassette if you have recorded one
3. Recorded audio cassettes or records for special music that you want to add live in the edit session
4. Artwork, photos, or graphics not already recorded that you want to add live in the edit
5. Credits or titles not already recorded that you want to add live in the edit
6. Voice-over announcer not already recorded
7. Other video recorded as part of another show or as stock footage

Using either the format, script, or storyboard as your plan for the video, you will know exactly how you want all these elements to fit together to make the show.

Viewing the Tape

Some think that you should sit down and view all the tape you recorded before you go into the edit. In essence, they are saying that you should not trust your logs but see for yourself exactly what is recorded on the tape. This is not a bad idea and if you have the time, by all means do it. How much time it will take you to do this is, of course, totally dependent on how much recorded material you have. It could be an all-day session or only an hour. It is a good investment in time and will help you with any lingering doubts about how to put the show together. If you have the luxury of doing this, you can check all the logs as you watch, making your own notes on the shots and creating the edit list right from the tape itself.

If you do not have the time to review the tape, you will have to rely solely on your shot logs. That is not bad either. After all, that is why you made them, isn't it? So you would not have to watch every piece of tape you recorded. If you have any doubts at all about what is on the tape, you can always selectively review the shots you have a doubt about, not the whole tape. That is the luxury you have when you have good shot logs.

The Edit List

The edit list is the show on paper, shot by shot, and is prepared on the form seen in Figure 12–1. Based on the format, script, or storyboard and using the shot logs, you create a list of all the shots you want to use in the show, in the order that you will use them.

If you view all the tape before going to edit, you can create this edit list at the same time. If you do not view the tape, rely on your shot logs, choosing your shots by those indicated as the best in the notes made by you and your crew.

An edit list does not lock you into anything. You can, of course, change and do something else while the edit is going on, if you wish. An edit list will, however, give you direction and save time. Without it, you will be shuttling the tape backward and forward, viewing shots over and over again. This can only lead to confusion with you rapidly losing track of what it is you are doing.

An edit list includes the camera number, the cassette number, the shot number, and a brief notation of shot content as follows:

Camera	Cassette	Shot #	Counter Number	Content Note
3	2	23–1	0296–0311	WS Reception

If you have more than one shot you want to consider (maybe you had two cameras recording the same action), view the tape for both of the shots and select one now while you are making the list. Do not wait for the edit to make the selection.

An edit list should also include an indication of any live taping you will be doing during the edit. For example, if you are recording a voice over live, you will include this in the edit list or if you want to roll in music direct from an audio cassette or record, you would indicate this as well.

Camera	Cassette	Shot #	Counter Number	Content Note
3	2	23–1	0296–0311	WS Reception
2	1	23–4	0197–0205	CU JOHN & MARY
LIVE	MUSIC UNDER	BEGINS HERE	AUDIO CASSETTE	
LIVE	VOICE OVER	BEGINS HERE	ANNOUNCER	
3	2	19–1	0156–0180	MS Band

Or if you are adding titles, graphics, or credits live, indicate this.

3	2	19–1	0156–0180	MS Band
LIVE	GRAPHICS	OVER SHOT #24–1		
2	1	24–1	0300–0311	MS J & M Dancing

Show Title _____

Date _____

EDIT LIST

CAMERA #	CASSETTE #	SHOT #	COUNTER NUMBERS	CONTENT NOTE

MATERIALS TO TAKE TO THE EDIT

Special art: _____

_____

_____

_____

End credits: _____

Title cards: _____

Figure 12–1   Edit List form.

The edit list, as we said before, *is the show*. Everything that is going to be in the show should be on it. Use the edit list along with the script (or format or storyboard) as your guide during the edit session.

HOW DO YOU USE THE EDIT TO MAKE A BETTER VIDEO?

A good edit can make even a bad idea into a good, even great, video. The success or failure of your video can be principally a factor of how good your edit is. Even if you do not have all the shots or if the ones that you do have are only mediocre, you can make a great video just by employing some tricks of the trade in the edit.

Your ability to make a better video with the edit is limited only by the equipment you have available to you and your own creativity in using it. If you have a full-blown edit suite like the big production houses in the broadcast television business have, you can do anything. You can make exciting video out of nothing just by virtue of the fact that you can create effects, even create complete pictures with this fancy equipment.

It is not likely you will have access to this though. It is more likely that you will have a punch and crunch system, that is cuts only, or that you will have access to limited effects through the camera, the VCR, or a small edit system. Even so, you can do a lot toward making a great video in the edit if you just remember you are still creating the show up until the time the edit master is complete.

First, you want to select the best pictures. The ones that tell the story best. Then you want to put these together in the best possible way. Here are some ideas.

Short Scenes

This may be the answer, particularly if you have a number of problem shots caused by the performers' not performing well or the visual not quite being right. Deal with this first by keeping to a minimum the time a shot is actually seen. Leave the shot on for as little as ten or even five seconds. This may sound like too short a time but in video, ten seconds on a shot that is not working can seem like an hour. Better to get off that shot before the audience has time to realize how bad or mediocre it really is.

By moving on to the next shot, you keep the action going and the audience involved in the story, not in analyzing how good or bad the shots are.

Shorten the Video

You may find that it might be best to change your plans entirely at least with regard to the length. A ten-minute show might make a better three-minute music video. A story that will get bogged down in ten minutes may be great shortened to three minutes. Set to a piece of music that captures the mood you want to create, the visuals take on a whole new meaning. The short cuts help. The music helps. The audience

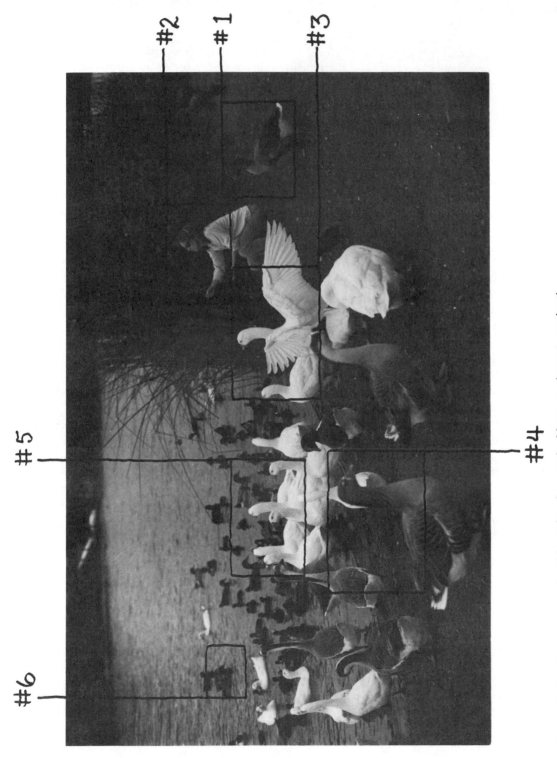

Figure 12–2 Photographs should not be shot static. Move on these to make the best use of the television medium.

helps because they get involved in the moment. And you have a dynamite music video rather than a boring ten-minute show.

It may even be exciting to create *two* shows, the long version and the short one. The wedding, for example, could be set to music for a short music video as well as the longer, more comprehensive twenty-minute show.

Be Flexible

The absolute worst thing you can do is to go into the edit session totally locked in. Such a rigid attitude will prevent you from adapting and changing the show as the edit requires. If you just did not get on tape what you thought you had, why not adjust instead of forcing yourself to use a bad shot? Throw the shot out or shorten the length of time it will be on. Or use something else entirely.

Remain flexible and open to different ways of accomplishing your objective. Just because you recorded a scene does not mean you have to use it. If it doesn't work leave it, as they say in the business, "on the cutting room floor" and completely out of the show.

On Shooting Photographs

Still photographs can add an exciting element to the video, enabling you to add visuals that are available only in stills. Use your imagination in shooting these, too. Just because you are taping a picture, that does not mean that you have to frame the entire picture, lock off, and tape static. Absolutely not. Photos will be more exciting if you let your camcorder do the movement for the picture.

For example, start on a closeup of someone or something in the picture and pull to a wide shot. Or cut from a closeup of someone or something in the photo to another closeup in the same still. Or push from a full frame to a closeup. The variety of shots available when taping a photograph is demonstrated in Figure 12–2. We can frame at position #4 or we could start there and push to #3 and then tilt up to #2 and pan to #1, finally completing the sequence by pulling out to #4 again. This would be much more visually exciting than just a wide shot the entire time.

The movements you can make on the photo are limited only by the composition of the photo itself. Do not, under any circumstances, just sit on a full shot of the photo for ten seconds. It will be deadly. And it will not be video. It will be a picture album and that is not what you are creating.

WHO GETS A CREDIT?

You as the producer and director decide who gets credit for their work on the show. We recommend that you always have credits. It makes the show look more professional and, besides, you owe it to all those great people both behind and in front of the cameras who helped to make it great. And it will make them all work just that much harder the next time.

You will have your credit list already started with the cast list and the crew list, so all you need to do is to add those not included there, put the names in the right order and put it on tape.

Credits are usually given in two separate lists, the cast first and then the crew. In both, people are listed in descending order of importance. For example, credits for the cast would be listed in this order:

> PRINCIPAL STARS
> SUPPORTING STARS
> ANNOUNCER
> EXTRAS
> GROUPS

You do not usually designate the role played by a cast member, only the performer's real name. Sometimes these names are put in alphabetical order if everyone is considered of equal importance or if you do not want to designate importance.

The crew credits would be listed in this order:

> EXECUTIVE PRODUCER
> PRODUCER
> DIRECTOR
> WRITER
> MUSIC
> LIGHTING DIRECTOR
> AUDIO DIRECTOR
> SET DESIGNER
> TECHNICAL DIRECTOR
> VIDEOTAPE OPERATOR
> CAMERA OPERATORS
> SET CONSTRUCTION
> GRAPHICS COORDINATOR
> MAKEUP
> HAIR
> WARDROBE
> GAFFER
> AUDIO ASSISTANT
> PRODUCTION ASSISTANT
> GO-FER

When doing crew credits, the crew member's title is followed by the name of the person as follows:

> EXECUTIVE PRODUCER
> James R. Caruso

DIRECTOR
Mavis E. Arthur

Another way to do credits is to combine cast and crew, beginning with the "big four" crew members, then to the cast and back again to the crew as follows:

EXECUTIVE PRODUCER
PRODUCER
DIRECTOR
WRITER
PRINCIPAL STARS
SUPPORTING STARS
ANNOUNCER
EXTRAS
GROUPS
MUSIC
LIGHTING DIRECTOR
AUDIO DIRECTOR
SET DESIGNER
TECHNICAL DIRECTOR
VIDEOTAPE OPERATOR
CAMERA OPERATORS
SET CONSTRUCTION
GRAPHICS COORDINATOR
MAKEUP
HAIR
WARDROBE
GAFFER
AUDIO ASSISTANT
PRODUCTION ASSISTANT
GO-FER

If you do not have time or do not wish to include all cast and crew in the credits, list only:

EXECUTIVE PRODUCER
PRODUCER
DIRECTOR
WRITER
PRINCIPAL PERFORMERS

As a part of your credits, you may also want to include some special *thank yous* to people or places that helped in the making of the video as follows:

Special Thanks To

The Boy Scouts of America
Saint Jude's Hospital
The Place Restaurant

This thank-you credit would follow the cast and crew credits.

Finally, you can give yourself a final plug as the show closes with a visual and voice over or a voice over only. This is the production company credit. The voice over might say:

THIS HAS BEEN A JAMES R. CARUSO PRODUCTION

If you have a logo or a special visual to go along with this voice over, use it. If not, at least take the audio credit. You deserve it.

ARE YOU READY TO DUB?

When you have finished the edit session and the show is on one piece of tape, this tape is your edit master. This is the original tape onto which you made the edit. Most material in it will be second generation. It is your number-one tape of the show and you therefore want to protect it from destruction. From it, you want to make one copy. This is your dub master. You will then store the edit master, taking it out only to make a new dub master or to reedit the show if you ever need to do either of these.

From the dub master, you will make your distribution dubs. You will not use the dub master as a viewing copy, only as a tape from which you can make dubs. The distribution dubs you make are for viewing. Make as many as you want from the dub master.

While videotape does not self-destruct over time like film, it will wear out or stretch from repeated play. It may even break or come off the cassette spools. You may begin to see some of this deterioration on videos you rent. This deterioration is not a major concern for you since you will not need thousands of plays for your videos. However, to be on the safe side, it is smart to start with the best tape you can buy. You will have done this with your original material. Now when you go to edit, you should make your edit master on the highest grade tape you can afford to lessen tape problems. Likewise, the dub master should be made on high quality tape.

The dubs are a different thing entirely. You may want to lessen the tape quality as a budget consideration for the distribution dubs. This will probably be fine. Most tape will record acceptable video and audio and if budget is a factor, you can feel safe in using this lesser quality tape.

CONCLUSION

You have now completed everything on your master checklist and in doing so, step by step you have learned how to plan a better video. To put all planning aspects into perspective in terms of how they are used, refer to the chart in Figure 12–3.

Note that in the preproduction stages of the show, you are using many planning techniques to help you to define the idea and develop this idea into a workable video. One by one you break the concept down into its parts and meticulously plan every part. All of these parts are then used on taping day for setup. However, once you begin taping at a location, you use only the script (format or storyboard), the shot list and the shot logs. All other lists prepared in the planning have served their purpose.

WHEN TO USE WHAT

	PRE PRODUCTION	PRODUCTION		POST PRODUCTION	
		setup	taping	plan	edit
Outline	X				
Budget	X				
Format script or storyboard	X	X	X	X	X
Schedule	X	X			
Equipment list	X	X			
Crew List	X	X			
Cast List	X	X			
Location list	X	X			
Prop list	X	X			
Shot list	X	X	X		
Shot log			X	X	
Edit list				X	X

Figure 12–3 Chart of planning steps demonstrating when they are used.

During the planning stages of postproduction, you need the script (format or storyboard), the shot logs and the edit list. In the actual edit session, you use only the edit list and the script (format or storyboard).

All planning elements have a specific function in making the video and once this purpose has been served, they are filed away in the show Bible for reference.

We encourage you to put the planning strategies outlined in this book to work by producing one of the scripts in Chapter 15, by making up your own based on the ideas in Chapter 14 or by producing one of your own ideas. We think you will find it both exciting and fun to see how easy it is to create great videos when you have a good plan.

That's a wrap on planning strategies.

Coming up next: *The Short Plan* . . . for those who are in a hurry.

TO DO

1. Make your Edit List.
2. Make your Credit List.
3. Make your dubs and distribute the show.
4. Check the Edit List, the Credit List, and dubs off your Master Checklist.

CHAPTER THIRTEEN

THE SHORT PLAN

Materials needed for this chapter:

Blank Forms

20-MINUTE PREPRODUCTION PLAN
15-MINUTE POSTPRODUCTION PLAN

If time is short and you just do not have the time to go through all the planning steps in detail, we offer here a short version that, while it will not assure you of total control over the shoot, at least will give you some control. It is designed to cover all aspects of the production cycle in extremely abbreviated form. And while it may be ludicrously short, certainly it is preferable to no plan at all. Included here is:

1. A 20-minute preproduction plan, shown in Figure 13–1
2. A production day checklist
3. A 15-minute edit plan, shown in Figure 13–2

Answer the questions on these forms and you can roll tape with at least an abbreviated plan. All times noted on these short forms are anticipated completion times, but we believe they are realistic. Should you wish to spend more time than designated, that is not only allowable but encouraged. Remember, the better the plan, the better the shoot.

While you are completing these short forms, if you want to time yourself just for fun, start your stopwatch at zero. If you are truly in a hurry, keep your answers short and concise to keep within the time limit.

Think fast. Ready? Go!

Show Title _____

Date _____

20-MINUTE PREPRODUCTION PLAN

TIME ALLOTTED	YOUR TIME	
00:30	_____	**WHAT DO YOU WANT TO DO?** (Briefly)

_____

| 03:30 | _____ | **WHERE WILL YOU TAPE?** List the locations then on a separate piece of paper, draw a basic sketch of each one. |

LOCATION

#1 _____

#2 _____

#3 _____

#4 _____

| 01:00 | _____ | **WHO IS IN THE SHOW?** List your cast as you now know it: principals, supporting, voice over, extras. |

ROLE   TALENT

_____   _____

_____   _____

_____   _____

_____   _____

_____   _____

| 01:00 | _____ | **WHAT IS YOUR EQUIPMENT AND CREW?** List the equipment and the crew assigned to it: cameras, production assistants, grips, gaffers, go-fers, and so on. |

Figure 13–1   20 Minute Preproduction Plan form.

EQUIPMENT   CREW

_____   _____

_____   _____

_____   _____

_____   _____

_____   _____

02:00   _____   WHERE WILL YOU PUT YOUR CAMERAS?
Using the sketches you drew for each location,
indicate exactly where you will put your cameras.

01:00   _____   WHERE WILL ANY REMOTE MICROPHONES BE
PLACED? Using the location sketches, indicate the
position of any microphones. Also list them here:

_____

_____

_____

01:00   _____   WHERE IS THE LIGHT AT THE LOCATIONS? Add to
location sketches the position of any lighting. Add to
this any lighting that you propose to take in. Also list
this additional lighting equipment here:

_____

_____

_____

03:00   _____   WHAT WILL THE SHOW LOOK LIKE? Using the form
provided here, draw storyboards for the three main
shots of the video. Make these simple but complete
enough to convey the idea.

Figure 13–1   (*cont.*)

06:00 _____ WHAT IS THE SHOT LIST? List every shot you can think of that you will want. Note the framing of these shots, such as CU, WS, MS, ECU, and MW.

SHOT #   FRAMING   DESCRIPTION

_____   _____   _____

_____   _____   _____

_____   _____   _____

_____   _____   _____

_____   _____   _____

_____   _____   _____

_____   _____   _____

_____   _____   _____

_____   _____   _____

_____   _____   _____

_____   _____   _____

Figure 13–1   (*cont.*)

01:00   _____   WHAT IS THE SCHEDULE? When and where will you tape? Day? Time? Place? And who is to report on those taping days?

PLACE _____

DAY _____

TIME _____

CREW CALLED _____

_____

_____

_____

CAST CALLED _____

_____

_____

_____

_____

_____

_____

Your plan is now complete. ROLL TAPE!

Figure 13–1   *(cont.)*

Show Title _____

Date _____

15-MINUTE POSTPRODUCTION PLAN

TIME ALLOTTED	YOUR TIME	
02:30	_____	ARE THERE OTHER THINGS TO SHOOT? Check your shot list to make sure you have recorded all shots. List any shots you still need, including graphics, photos, cutaways, and the like.

SHOT #   FRAMING   DESCRIPTION

_____   _____   _____

_____   _____   _____

_____   _____   _____

_____   _____   _____

_____   _____   _____

01:00	_____	ARE THERE ANY TITLES OR CREDITS? Write down the name of the show and any credit you want to include. Do not write in specific names for cast or crew; you have these already listed on your 20-minute preproduction plan. Note only that you want to include cast and crew in the credits, if you do.

TITLE _____

CREDITS: Cast _____   Crew _____

Others _____

_____

_____

01:00	_____	DO YOU WANT TO ADD ANY ADDITIONAL AUDIO? Music? Announcer?

_____

_____

Figure 13–2   15 Minute Post Production Plan form.

_____
_____
_____
_____
_____

02:00 _____ WHAT EFFECTS DO YOU WANT TO ADD IN THE EDIT? Are there any special effects, audio or video, that you want to add?

AUDIO EFFECTS

DESCRIBE _____
_____
_____

VIDEO EFFECTS

DESCRIBE _____
_____
_____

08:30 _____ MAKE THE EDIT LIST. Using the shot list, make an edit list in the exact order that you want the shots cut together in the final show.

CAMERA   CASSETTE   SHOT   COUNTER NUMBER

_____   _____   ____   _____
_____   _____   ____   _____
_____   _____   ____   _____
_____   _____   ____   _____
_____   _____   ____   _____
_____   _____   ____   _____
_____   _____   ____   _____
_____   _____   ____   _____

You are now ready to edit.

Figure 13–2   (*cont.*)

THE 20-MINUTE PREPRODUCTION PLAN

<u>TIME ALLOTTED</u>

00:30   WHAT DO YOU WANT TO DO?

Very briefly answer this question. Your answer might be:

<u>Tape a western shootout</u>

03:30   WHERE WILL YOU TAPE?

Make a list of your locations. Be specific, making sure to list every location. Once this list is complete, make a basic sketch of each location. You will need this later. Our examples are shown in Figure 13–3.

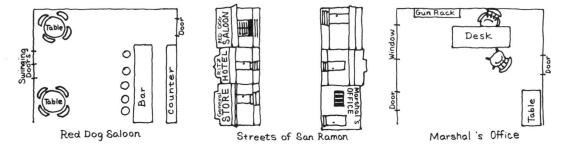

Figure 13–3   A basic sketch should be drawn of every location.

1.   Marshal's office
2.   Streets of San Ramon
3.   Red Dog Saloon

01:00   WHO IS IN THE SHOW?

Who are your performers? List them by their real name and by their name in the video, if it is different. Be sure to cast all roles. For crowds or audience, note only that they are there. Do not list them by name.

Luke—Joe Smith
John Sleeze—Jonathon Mack
Mary Sunshine—Sue Wells
Townspeople

01:00 **WHAT IS YOUR EQUIPMENT AND CREW?**

List the equipment you will be using and the crew member assigned to that equipment. Even if you are taping a one camera shoot and you are the camera operator, you should at least have an assistant to keep your taping logs.

Camera #1 (tripod)—Harvey Johns
Camera #2 (handheld)—Marie Claar
Production assistant—Virginia Hoover
Production assistant—Kathy Groza

02:00 **WHERE WILL YOU PUT YOUR CAMERAS?**

Using your location sketches, indicate exactly where you will put your cameras, as demonstrated in Figure 13–4.

Red Dog Saloon   Streets of San Ramon   Marshal's Office

Figure 13–4 Cameras are drawn into the sketches in their taping positions.

01:00 **WHERE WILL ANY REMOTE MICROPHONES BE PLACED?**

Using the location sketch, indicate the position of any microphones, demonstrated in Figure 13–5. Also list them here:

Boom microphone—Over action
Remote microphone—Handheld

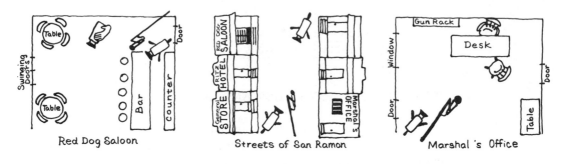

Figure 13–5 Audio placement is noted on locations sketches.

01:00   **WHERE IS THE LIGHT AT THE LOCATIONS?**

Add to your location sketch, as shown in Figure 13–6, the location of any lighting that you may be aware of. If you do not know the answer to this question, you will have to wing it at the location.

Figure 13–6 The placement of any lighting is indicated on the sketches.

03:00   **WHAT WILL THE SHOW LOOK LIKE?**

Sketch the three main shots of the video on a storyboard form. Make these storyboards simple but complete enough for you and your crew to know exactly what you have planned, as demonstrated in Figure 13–7.

06:00   **WHAT IS THE SHOT LIST?**

List every shot that you can think of. Since you will not have extensive plans, this shot list will note only basic shots. You will want to add to this the closeups and cutaways as you tape. Group these shots by location.

LOCATION: Red Dog Saloon
E = Exterior   I = Interior

CONFRONTATION IN THE RED DOG SALOON.	SHOOT OUT ON THE STREETS OF SAN RAMON. SLEEZE FACES THE MARSHAL AND JERRY LEE.	THE MARSHAL IS CONFRONTED BY THE TOWNFOLK WHO TRY TO GET HIM TO BACK DOWN TO SAVE THE PEACE.

Figure 13–7 Draw three panels of a storyboard for your show, to help in the visualization.

E   #01: WIDE SHOT, PUSH TO CU OF ENTRANCE DOOR

I   #02: WS OF THE BAR AREA

I   #03: MS OF TABLE TWO FRIENDS DRINKING

I   #05: CU SLEEZE

01:00   WHAT IS THE SCHEDULE?

When and where are you going to tape? Day? Time? Place? And who is to report that day? Crew? Cast?

Place: Red Dog Saloon
Day: Monday, June 2
Time: 9 AM
Crew: All
Cast: All

Your Preproduction Plan is Now Complete.

In addition to these questions, ask yourself, "Is there any dialogue I particularly want?" We have not included this in the twenty minutes for obvious reasons. Dialogue takes thought and time to get it just right. We encourage you to spend the extra time if your video calls for scripted dialogue or if a scripted piece will make the video better.

Write down any dialogue you want said. If you are not sure of the exact words but you do know the basics, write them down.

WHERE SAID: Bar at the Cattlemen's Restaurant

<u>BAD GUY</u>

Get out of town by noon or I'll gun you down like a dog.

<u>GOOD GUY</u>

This ain't your town. It's mine and I ain't leavin', not for you or any of your kind.

<u>BAD GUY</u>

Then get your gun, lawman. I'm comin' after you at noon.

Make copies of this plan and distribute it to all cast and crew. This will become your taping Bible, answering questions about the show, the schedule, and what is expected of everyone.

A PRODUCTION DAY CHECKLIST

On the production day, you want to follow a certain sequence in order to make sure that the taping goes smoothly and that you are prepared for the edit. Have your 20-minute preproduction plan available to assist you in going over the following checklist. Follow the steps in this order:

On the Shot Day, Before You Roll Tape

1. EQUIPMENT AND CREW CHECK.   As they arrive at the location, check off equipment and crew listed on the preproduction plan.
2. CAST CHECK.   As they arrive, check off your cast listed on the preproduction plan.
3. POSITION THE CAMERAS, AUDIO, AND LIGHTING.   Using the sketch made during the preproduction plan, position the equipment and assign your crew their positions.
4. WARDROBE AND MAKEUP CHECK.   Check your cast for wardrobing and makeup.

5. SHOT LIST ADDITIONS.   Add to the shot list, made in your preproduction plan, if the circumstances warrant it.

When You Are Taping

6. SHOT LOG.   Log all your shots on a shot log. Include on this shot log every shot recorded on tape, including those on your shot list and those you add, such as cutaways.

Before You Wrap The Shoot

7. SHOT LIST CHECK.   Check your shot list to make sure that you have recorded all planned shots.
8. FINAL CHECK.   Ask yourself, "Did I get everything?"

THE 15-MINUTE POST PRODUCTION PLAN

After you have finished taping all your shots, before you begin the edit, take 15 minutes to make an edit plan. Use your shot log and your 20-minute preproduction plan to help answer these questions:

TIME ALLOTTED

02:30   ARE THERE OTHER THINGS TO SHOOT?

Check your shot list again to make sure you have recorded all the shots. Since you have no format, script, or storyboard, run through the show as you see it in your head. Do you need to add any graphics or photos? List these additional shots. Number them in consecutive order from those you have already recorded.

SHOT FRAMING AND DESCRIPTION

#025   WS PHOTOGRAPH OF WESTERN TOWN TO CU SALOON

#026   ECU TO FULL OF CLOCK TICKING TOWARDS NOON

#027   WS PHOTOGRAPH OF SUNSET TO RACK FOCUS

01:00   ARE THERE ANY TITLES OR CREDITS?

Write down the name of the show and any credits you want to include. Do not write in cast or crew here. You already have that list on your 20-minute preproduction plan. Just make a note if you do want to include these.

TITLE: SHOOTOUT

 CREDITS

Cast   X   Crew   X

Other   PRODUCER—Mary Jones

 Troup #34 Boy Scouts of America

01:00   DO YOU WANT TO ADD ANY ADDITIONAL AUDIO?

Music? Announcer? Sound effects?

MUSIC: THEME FROM THE MOVIE "HIGH NOON"
OPENING VOICE OVER —"This is San Ramon, a town that is
about to have a problem, a problem with time and bullets."

02:00   WHAT EFFECTS DO YOU WANT TO ADD IN THE EDIT?

Are there any special effects you want to create in the edit session? This list will probably be short since your capabilities will be limited. Use your imagination. There are lots of things you can do without all the fancy equipment.

 VIDEO EFX: RACK FOCUS ON SUNSET

 AUDIO EFX: MAKE TICKING OF CLOCK A BOOMING SOUND

08:30   MAKE THE EDIT LIST.

Using the shot list, make an edit list, the order of the shots as they will be cut together in the show.

Camera	*Cassette*	*Shot*	*Counter Number*
1	3	025	0026–0087
2	1	002	0127–0156
2	1	003	0200–0230
1	2	010	0110–0132

YOU ARE READY TO EDIT.

While these short cuts to planning are not meant to be extensive, they are fairly comprehensive and by using them you will have a more structured shoot.

Coming up next: *101 Ideas for Videos* . . . to get you started.

CHAPTER FOURTEEN

101 IDEAS FOR VIDEOS

We know that you have lots of ideas for videos. Everyone does. We want to encourage you to think further though, beyond the everyday, beyond the family and friends, even beyond what you see on television or at the movies. Think creatively. Make your own personal mark on the video world. Create some truly unique, and maybe even outrageous, videos of your very own. Here's a list to get your thinking process started.

A Day in the Life of . . .

Pick your subject and spend a day with him or her, taping their everyday activities. Ideally this video would begin at the beginning, sunrise and waking up, and proceed to the end, sunset and falling asleep in front of the TV or whatever.

A Letter Home

Instead of writing a letter, why not deliver it in person on videotape? You can talk directly to the camera as if you are talking to your loved one. Intercut shots of what you are talking about, like the place you live, where you work, your latest A at college and even some of your friends. You could do this on a regular basis or once a year at Christmas to let friends and family know what you have been doing the past year.

A Night Out

Getting out to the movies, dinner, or the theatre may not be something you can do every day, so why not record one of these occasions? Then when you need a night out but just cannot break away, you can view this tape and recapture the feeling.

A Summer Night

A collection of summer nights, including sunsets, fires on the beach or lakeside, and sunburnt bodies in bathing suits. This tape will be a joy in December.

After Midnight

What happens after midnight in your city? In your life? Take one month in your life and tape what you are doing at midnight during that time. Edit these together and you have a good representation of what happens to you After Midnight. You might find the results interesting.

America

Or Americans, if you prefer. An up-close look at your country, the people that live in it, or both. Start with your own surroundings and then build on them with news, movie, and television-show footage.

Animated Short

The principle of animation is: Record something that does not move, such as a picture or clay figure. Stop tape. Change still picture or object to the next movement. Record another shot. And so on until all shots are recorded. Then to animate the action, simply cut one second of shot #1 to one second of shot #2 to one second of shot #3 and so on. By cutting shot to shot in short sequences, the characters appear to move. You can also make an animated short with toys that bend or with clay figures. You simply shoot the characters, change their position, and shoot them again. Then shoot, change, shoot, and so on. It is tedious but rewarding. Look what they did with the California raisins! Good luck.

Applause! Applause!

Give yourself a hand. Record people cheering and applauding and encouraging you to greatness. When you need it most, you will have this videotape to push on toward accomplishing all your dreams.

Baby Firsts

A new baby is always a good subject for numerous videos, including *Learning to Walk, The First Year, First Christmas, First Haircut, First Words,* and *Learning To Eat*. These videos may take time, being shot over a rather long period of time, so be sure to keep good logs so you can find those great shots again when you get around to editing the material. You are likely to have tons of material from which to choose but it should be fun to put together and great to watch.

The Beach

Choose your best beach, set up your camera at the angle you love best, and let the tape roll (Figure 14–1). You do not even have to move the camera unless you just want to. The gentle roll of the ocean, the breeze fluttering the palm trees, and

Figure 14–1 *The Beach* video.

you quietly walking down the beach may be enough. Do not edit this tape. Let it run au naturel, an hour or more. Play it when you just cannot take it any more. It will help you keep life in the right perspective.

Be My Valentine
Make a video valentine for your sweetheart. Include all the sweetness and love you want to express. Use flowers and candy, lots of kisses, and of course *you*.

Beauty Contest
Stage your own beauty contest. There are two ways to do this. Star your friends wearing their best and most gorgeous swimsuits. Or take one or two of your friends to your local department store. Have them try on different swimsuits. You can even carry this on to another store. You may find this a great way to buy a swimsuit. Put it on, tape it, and take it home on video *before* you choose. The sales person at the department store should be impressed by your imaginative approach and more than willing to cooperate.

Best or Worst on TV
Use footage from the TV shows themselves. Do your own voice-over narration at the appropriate spots to comment on the wonderfulness or awfulness of every show. You might want to limit this to the ten best or worst. Otherwise this video could go on forever. You could make this a yearly tape. Document the kinds of shows on the air every year while documenting your own likes and dislikes.

The Best Things in the World

What exactly do you think are the best of all possible things in the world? They can be anything, anyone, or anyplace. Choose your best of the best and document them on videotape. You could also do the Worst of Things version if that is to your liking.

Birthday

Tape anyone's birthday. Yours. Theirs. Or whoever. Or maybe a surprise birthday party. Or a special birthday such as Sweet 16 or Finally 21.

Bloopers

Make your own blooper show with outtakes from your own shows. Outtakes are those mistakes, those screw ups, those cute little unexpected things that happen when you are taping. They make for a really funny video.

Bridges

Get in your car and drive over different bridges with tape rolling. The visual you will record is almost like a special effect. Every bridge will be different. The movement. The surroundings. You will find the effect interesting.

Circus

Make your own circus. Everyone has some trick they can do, even if it is a "dumb" trick. Maybe you have a cat that fetches or a dog that catches Frisbies™ or maybe you can catch a Frisbie™. Turn these amazing feats into a spectacular "greatest show on Earth."

Christmas

A special Christmas or a year-by-year account of Christmases. Some Christmases are better than others. But they are all worth remembering. Cut in Santa Claus. Use pictures from a past Christmas. You can have anyone home for Christmas when you have them on your Christmas tape.

Car Chase Scene

You have seen these a million times on your television—one car chases another, avoiding crashes with daring maneuvers, only to end up in a spectacular crash. You can do this. You can create your own with toy cars or real cars in fast-forward with shots or you can use part of your favorite car chase, all intercut with shots of you at the wheel of a similar car. Even the sound effects here should be easy, taken either from a recorded show or created with your own voice or car.

City I Live In

Document the city you call home. It has its own personality and part of that is reflected in your lifestyle. Look at the city as a tourist would. Visit all those spots

Figure 14–2   *The City I Live In* video.

everyone has seen but you. Isn't it time you took a good look at where you live? (Figure 14–2)

Cowboys and Indians
Remember when you used to play cowboys and Indians as a kid? Play it again or let your own kids play for you. They will have their own way of playing, but maybe you can convince them to play your way. You could intercut shots of yourself as a kid, putting yourself into the game just like it used to be.

Dad
Dad deserves a tape. Where would you be without him? Set this video to music that makes you all emotional and gushy. Use snapshots from the past intercut with Dad today and, of course, you with good old Dad.

Dancing
You should be dancing! The two-step. The samba. Rock and roll. Or just let it all go where it wants to to the beat of the music. Dirty or clean dancing. Or a little of both. This ought to be *hot!*

Dreams
Had a great dream lately? Why not act it out on video? Could be good therapy, even fun, depending on the dream. Good or bad, it probably will *not* be boring.

Exercise

What is your game? Tennis? Running? Aerobics? Swimming? Whatever it is, what do you look like doing it? Do you really want to know? Why not? You are probably gorgeous, so why not capture the real you on videotape so you can show the world how hard you work to stay slim and beautiful?

Family Court

Don't get even with those relatives that have done you wrong. Take 'em to court. Do it tongue-in-cheek, but focus on real problems. Maybe, with humor, you can reach a solution and make a unique family video at the same time. Don't forget the bailiff. You might need one before this tape is over.

Family Tree

The Who's Who of your own family. Trace your family tree on video using photos, film, video, or a drawing. Start with the present and work backward as far as you can go. You could even speculate on your relatives 2000 years ago or even further back.

Fantasies

Indulge your fantasies in video. You can take that trip to the moon, drive that Ferrari, or fight Darth Vader. You can lie on a warm beach, be a movie star, or be at a party with your favorite celebrities. Anything is possible in video. All you have to do is choose your fantasy, gather the prerecorded video or tape some new material, and you are there.

Fat

Putting on weight? Taking it off? Or are you just interested in *fat?* Maybe you can document the beginning of a great new figure and the work it took to get there. A before and after on video.

Feet

Have you ever really looked at feet? There are all different kinds. Big. Small. Long toes. Short toes. Flat feet. Arched feet. Record only feet for this video. Bare feet. Feet with shoes. Feet with socks. Have the feet wave at you. Walk for you. Run for you. And do all those things that feet do. You will find when this is edited that there are more kinds of feet than you ever imagined. Once you have this show finished, it could become a quiz for your friends and relatives. Identify the owner of the feet shown. Can you find yours? (Figure 14–3)

Fireplace

Tape your own video fireplace. Your own yule log. Get the fire ready to light. Set up the camera so that only the fireplace shows in the frame. Roll tape. Light the fire. You may want to push to a closeup once in a while but mainly you will just be rolling tape on the locked-off shot of the fireplace burning. An hour or so of this and

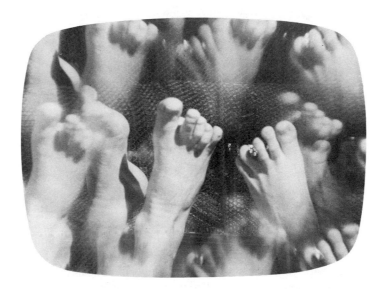

Figure 14–3 *Feet* video. Effects Filter courtesy Cokin® Filters.

you will have a video you can play when you do not have a fireplace or when you just do not want to light the one you have. By the way, don't forget the audio. (Note the specifications on your camcorder regarding light and heat before taping this video.)

First Day of School
The jitters. The crying. The hand holding. All those preparations for school including buying the right clothes, school supplies, and perhaps a nostalgic look back at your own school days.

Food Fight
Stage your own food fight. Be sure to put more than one camera on this to capture all the action as it happens. You do not want to have to clean up this mess twice.

Fourth of July
Fireworks. These are not easy to tape, but spectacular when you get them. If you just lock off your camera for ten minutes or so in the area where they are going off, you will get good base. Then you can try to anticipate and get some tighter shots of various explosions. Intercut this spectacular display with kids eating hot dogs, Mom cutting her homemade apple pie, and Dad playing ball with the kids. You will have on video an old-fashioned Fourth.

Freeze Frame
Look at the world frozen in time, frozen at those exact moments you love the best. A sunset at just the right moment. The one you love looking at you in just the

right way. Your favorite star captured doing what you love the best. Make this video by taking the recorded tape and letting it roll up to *that moment*, then freeze frame and hold it.

Frogs

Frogs are lucky, you know. Be kind to a frog, be a success in the world. Well, it's an opinion. Maybe you think owls are lucky. Or mice. Or cats. Or dogs. Whatever. Make a video of what you think is lucky. Then when you need the luck, you'll be ready with the video.

Game Show

If you have always thought you could play your favorite game show better than any contestants you have seen on the show, now is your chance to prove it. Tape your own version of this game show or any other game show. All you need to do is tune in to the show, outline the format (game shows work from a basic format rather than from a structured script), get the props you will need and go to it. There are even some board-game versions of these shows in your local toy store that will make this even easier to do. If you roll tape on the show itself, you can even have the same prizes and the same prize girl in your video that they have on the regular show. Even the regular host can be cut in to make his cute remarks.

Going Home

Can you go home again? Why not? Begin with photos of the way it used to be. Then go home and see the way it is today. Back to the old school. The house you lived in. The friends you knew. This should be a nostalgic journey.

Gorgeous Women/Men

You can include your own lovely lady or man in this video. Or your friends or relatives. Use pinup posters or calendars. Or take a shot from a commercial or a television show or a movie. A video of gorgeousness. What more could you ask?

Growing Up

This could be about you, your wife, your kids or anyone else. You will have to draw on old snapshots and video for this video. Do not forget to cut in news stories or TV shows from different eras. And, of course, back it all up with music from the different time periods. If you grew up in the '50s, for example, you would want to include Elvis Presley, Hula Hoops℠, sock hops, and '57 Chevys. These are all available on video if you just use your imagination in locating them.

Having a Baby

This is more of a video about Mother and Father than it is about Baby, since it is Mom and Dad that you will be recording as they prepare for the arrival. You could include getting the nursery ready and, of course, Mother's growing tummy.

You could end with the arrival of the little one or you could take the baby home for the big finish.

How to Make Your Favorite Recipe

Share the secret of your favorite recipe. Shoot this like a cooking show with you as the chef. Use closeups on the food so we can see all ingredients and how to mix them. Stand on a chair and shoot down to see inside a pan or bowl. Be sure to include specific ingredients and measurements. This is serious stuff. (Figure 14–4)

Figure 14–4 *How to Make Your Favorite Recipe* video.

I Hate These Things

A video montage of what you hate most in life. Things. Foods. Places. Jobs. People. Events. Whatever. Let it all hang out and get it out of your system.

I Love These Things

What do you love? People. Things. Food. Clothes. Places. Sports. Whatever. Love them on video for everyone to see. Careful—keep it PG.

I'd Rather Be . . .

What would you rather be doing? Sailing? Swimming? Skiing? Lying on the beach? This video can put you right where you want to be, anytime. Just go where you would rather be and roll tape. If this is not possible, use photos you have taken or found in, say, a magazine. Use video from commercials, TV shows, or movies. If you would rather be sailing, then sail vicariously through this video.

If I Had a Million Dollars

Indulge yourself. What would you spend your million on? Tape shots of the house you would buy, the car you would drive, the vacation you would take, the diamonds you would wear, and the party you would throw for your celebrity friends. You can tape most of this yourself. It will be fun to shop, to imagine, to fantasize, and perhaps to even feel like it is real by having it on tape.

If I Won the Lottery

This is a great fantasy. Say the prize is $12 million dollars or even $40 million. You can pick your own amount. Make it $100 million. You can do anything. Go anywhere. Let your imagination run wild. Then do your video shopping. Capture it all on tape and the fantasy will be yours whenever you like.

It's a Jungle Out There

Life is full of lions and tigers and bears. Can you survive? Sometimes it doesn't seem likely, but inevitably you do. Maybe if you rolled tape on the jungle, you could cope with it better or maybe even learn to laugh at it.

Jokes

Tell your favorite jokes on camera. This could be your beginning as a standup comedian. No cutaways here. You are on your own. Smash pumpkins, wear funny hats, use squirt guns. Choose your weapons and roll tape on jokes. (Figure 14–5)

Figure 14–5 *Jokes* video.

Kids Say the Darndest Things

Kids at a certain age are priceless. Capture that on video with closeups of them defining the world for you. You might, for example, ask them what Christmas means or what the Star Wars Initiative is or how to make peanut butter and jelly sandwiches. The questions are endless. The answers are totally unpredictable.

Late Night With . . .

Have your own late-night show. Who would be your host? Who would you interview? What would you ask them? This is a simple video with little or no editing. It is a one-on-one interview situation. May the best man (or woman) win.

Life in the Fast Lane

Run the world in fast-forward. It works better that way sometimes. Record the action and then speed it up. This will always give you a laugh.

Lifestyles of the . . .

What? Rich and famous? Poor and everyday? Average and boring? Where do they go? What do they do? Do they have special hideaways? Special hobbies? Special taste in food? What do you think? Or do you even care?

Loving

You can feature your own loving experience with your wife or husband, girlfriend or boyfriend, kids, parents, or special friends. There are all kinds of loving. You can feature them all or choose one and focus on that. A liberal use of closeups, soft focus, and slow motion, backed by the right music, will make this an emotional high.

Loving TV Style

Here's your chance to create your own version of the way they love on TV. Daytime soaps are a good source of ideas for this one. Then, of course, there is the late-night version of love, which involves a lot of teasing but little delivery. You might want to show it the way it was on TV, then follow this with your version of that same scene. Could be interesting and fun.

New Year's Eve Resolutions

All right, here is the ultimate challenge. Put those New Year's Resolutions on tape. Then there will be no way you cannot live up to them. Tape not only your vocal promise but a visual representation of that resolution. For example, if you are going to promise to lose ten pounds, show us the fat and the food that put it there. You are going to love this tape a year later. It will come back to haunt you.

Night Streets

The lights of the city, the cars on the freeway, your own house lit up. You will have to push your camera to the limits lightwise to get this one. You may not even be able to get it clearly enough with so little light. But give it a try.

Me

If anyone or anything deserves a video, it is you. So go for it. You are the star! You do not even have to include other people unless you want to. You can always set up the camera on a tripod or some flat surface and do whatever you want for your own personal video. It would be cute to add to this tape yearly to show the different faces and personalities of *you*.

Mom

Okay, Dad has his. Here's Mom's equal time. After all, who could be more important than Mom? No one. She certainly deserves a video of her own. A nice music video set to emotional, I-love-my-Mom type music would be perfect.

Monsters in the House

There are monsters in your house, whether you know it or not. There are bugs of all kinds everywhere. Shoot these in closeup and they become monsters. Monsters such as these have been terrifying entire cities on the movie screen for years. Now you can capture the monsters in your house in this video. Just tape every monster in closeup as they crawl by or as they attack each other. A spider with a fly captured in its web is gruesome in closeup. Potato bugs are pretty awful, too. (Figure 14–6) Watch out that the monsters don't get *you* while you are making them stars.

Figure 14–6 *Monsters in the House* video.

Moving to Our New House

First of all, you will want to tape the new house empty before the move-in. Then tape the packing and all the craziness that goes along with it, the actual moving day ending with the first meal prepared and served surrounded by the chaos of the move. Finally, you might want to end with a shot of the house a month after the move-in to show how you have settled in and made it into your own special home.

Mud or Oil Wrestling

This ought to be fun. You do not have to be a professional to do this. Find a fun-loving friend, run some water over dirt for an hour or two and then go to it. Set the rules that the first one thrown from the mudpit loses. Or two out of three if you want the fun to last longer. If you do not have any mud, do it with baby oil. Rub down liberally with the oil. Tack a sheet down to the floor using thumbtacks or staples or nails or whatever. Or make the whole room the wrestling arena. First one out of bounds loses. Be sure to put someone really good on the camera. You do not want to miss any of this action.

Murder Mystery

Everyone loves a mystery. So make your own. You can buy murder mystery games based on video and you could pattern your video after these. Or you could just create your own version. All you have to do is stage a murder, introduce the possible suspects, and then supply clues to solving the puzzle.

My Blood or Yours

Vampires. In the light of the full moon, the vampire walks, hungry for blood. Tape the full moon, ideally with clouds drifting by. This makes it more ominous. Make your fake blood so it can drip realistically from the vampire's fangs. Cast your sweet virginal maiden and do not forget the all important black cape and the hero, of course.

My Car

Everyone loves their car so why not keep it forever, on tape. This could become a documentary of all the cars you own with shots of them when they are new and when you trade them in on another one. You may even want to include shots taken while driving, showing you at the wheel. This is a lifetime project unless you put some limits on it.

My Favorite Food

What kind of food do you like most? Why not put it to music, as they say. Maybe it is a special kind of candy. If the factory is close by, you can go tape shots of them making it. If it is not nearby, the factory may have some tape they would be willing to loan to you. You can roll over a dub and return their tape. Follow the same procedure for any other kind of food. If nothing else, you can cut various shots of you buying, preparing, and eating it, set to the appropriate music.

My Favorite Person

Intercut video with photographs; set these to music that is appropriate to the way you feel about this person. This can be a close friend or family member or it can be someone you admire for their talent, like Michael Jackson, Madonna, or Frank Sinatra. If it is a well known person, it will be fairly easy for you to find prerecorded video either in your video store or airing on broadcast or cable television.

My Favorite Year

They made a movie out of this. It was a funny year in the movie. Yours may or may not be funny but it should be memorable. You can document this year with video you might have taken, photos, graphics, and footage from movies and television shows that were on the air then. It will be interesting for you to have to go through the exercise of choosing your favorite year, don't you think?

My Hairdos

Every time you get a new hairdo, roll tape. Record every new hairdo that you get for, say, a year. Then sit down and take a look. You will have the opportunity to take a look at yourself from a new perspective. The perfect hairdo for you will probably pop out at you and you can create exactly the image you want.

My Baby Laughs and Cries

Over a period of several months, make an effort to record your baby laughing and crying. You will discover that he or she has different laughs and different cries. This will be a really cute video that will become a family classic.

On the Job

Get someone to follow you one day, taping your daily work activities. You could begin this video with the happy tune "Hi Ho, Hi Ho, It's Off to Work We Go" and by the end of the day you may feel like closing it with "Take This Job and Shove It." (Figure 14–7)

On Vacation

This could be an ongoing series with you adding a three-minute montage of your vacation every year. It might be interesting to see how your likes and dislikes change over the years. It could also be a video about just one particular special vacation.

Pet Peeves

What bugs you? Put it on video and you can let others know without offending them. This tape should be humorous and fun to watch. For example, "I hate it when you cough at the dinner table. I hate it when it rains and I just washed my car. I hate it when the phone rings and you just sit there and wait for me to answer it." This might turn out to be video therapy for you and your subjects.

Figure 14–7 *On the Job* video.

Predictions

Make your own predictions about yourself, your friends, your relatives, the world, or certain movie and TV stars. What will happen in the next year? You decide. A year from now you can edit to this what really happened to see how accurate you were.

Reunion

The Class Reunion. A Family Reunion. A gathering of friends. Intercut the last reunion or footage of the era the reunion commemorates. The Class of '88 could include news pieces, movie clips, and TV shows.

Rubber Duckie

Rubber duckies bouncing along in the bathtub, squeaking when you gently swish them. This is a short video set to rubber duckie music. A rubber duckie music video. Sounds great.

Send Money

This might be your most important production. How would you convince someone to send you money? With humor? With shots of you begging for food on the street? Would you offer them something in exchange? Would you get tough? Or sweet? Let your imagination go. Be outrageous. Get their attention. But get the money.

Seniors

A loving look at the older people in your life. Their hopes and desires. Their plans for the future. What do they mean to you? Use soft light and soft focus to make them all look younger than they are. They'll love you for it.

Smile

Make 'em smile for you. Everyone you know and some you don't. A show of smiles will be a great "upper" to watch when you are feeling a little low. Just smiles from the whole world. (Figure 14–8)

Figure 14–8 *Smile* video.

So What?

Do you ever get tired of the same old news about the same old people? Do you ever get fed up with being told the same old thing by the same old people? Do you ever want to say *SO WHAT!?* Say it in a video. Intercut *SO WHAT!* with video of the so-whats in your life. You might start with just you saying *SO WHAT!* after the first clip. Then after the second maybe you will be joined by one of your friends and you will say *SO WHAT!* together. Continue doing this until there are thousands (at least ten or so) shouting *SO WHAT!* into the camera.

Soap Opera

Tape your own soap opera. You can write a totally original script or you can tape a soap off the air, write the script by viewing the tape, and then shoot with your own talent. This ought to be fun, perhaps even dangerous.

Special Movie Scenes

Recreate your favorite movie scenes. From *What a dump* to *Here's lookin' at you, kid* to *if you want me, just whistle*. Or the scene where Indiana Jones throws down the sword and shoots the Ninja warrior. Or where Gary Cooper takes off his badge and throws it into the dirt in "High Noon." There are classic scenes that we can never forget. Make your own version using the original as your script.

Sports

What is your sport? This could be a video about your favorite football team using actual footage from the game or it could be a video about your son's little league baseball team. Or it could be you showing us your expertise on skis. Choose a sport and put it to tape.

Storming

Everyone loves a good storm. Wind blowing. Snow or rain. An angry ocean. A roaring fireplace. This can be a video about one really good storm or you can capture on tape different aspects of different storms and then edit these together. The crashing of the surf against the rocks. Rain pounding on the windshield of your car. The wind whistling through the walls. This could be a video you want to watch in the middle of summer when the weather is perfect but you could just use a good storm.

Strip Poker

Rate this one PG, R or XXX. The choice is yours and so is the style of the video. We leave that to your imagination.

Sword Fight

Swashbuckling. Defending my lady's honor with a sword. Buy toy swords but make sure they are the long ones, so it looks authentic. Choose two swordsmen and let them have at it. The action should be intense. Be sure your camera operator can keep up.

That Was the Way It Was

Choose a year, a day, or a week and put on video what happened that day, week, or year. Include national events as well as your own personal life.

These People Should Get Out of Town

This is your golden opportunity to express your feelings in video. Let it all go and include all those people you could live without. At least you could live better with them in another town. Use closeups and footage of what irritates you most about these people. Let 'em have it.

Top Ten Best/Worst-Dressed

Make yourself the expert. Make your own best- or worst-dressed list. Find video or photographs of likely candidates and then choose your own ten. Pick on your own family, friends, fellow employees, or national stars. You be the judge. Be sure to put the year on this video. You may want to do it again or make it a yearly project.

Top Ten Movies

Again, be sure to put the year on this video. Draw your material from the movies themselves. Use clips to demonstrate why you think they are the best or the worst. You could even set yourself up as on-camera narrator, like a movie reviewer. If you think about it long enough, you can probably even find a theater where you can tape your sequence. Try the local high school, or the local movie theater might even let you come in and tape it. If not, try the local community theater. And don't forget the popcorn.

Treasure Hunt

Create your own treasure hunt. Hide the treasure, make treasure maps, divide your players up into teams, and let them go. This will be just as much fun watching as it is to make.

Trekkies

If you are a *Star Trek* fan, you have this one already figured out. You can draw from the old television show, the new television show, the movies, the clubs, and conventions held nationally.

Trivia Quiz

Use video to ask the questions. Choose a subject like current events, commercials, movies, or television and use video from these sources in this unique Trivia Quiz. You might, for example, ask the name of the star of *Dallas,* showing his picture at the same time. This is not a video you watch. Rather, it is one you play with.

Video Aquarium

Set up your camera so that only the aquarium shows in the frame. Lock off. Roll tape. You may want to push to a closeup of a fish now and then but mainly this video will be a wide shot of the fish swimming lazily in the aquarium. Record an hour or so of this, and *presto!* You have an aquarium whenever you like. Just play the tape.

Video Pet

Have your own pet on video. No mess. No fuss. Choose your pet. Something conventional like a cat or a dog. Or something more exotic like a gorilla or a tiger. You can have any pet you want. Borrow your friend's pet cat or go to the zoo to tape your pet. Concentrate on closeups. (Figure 14–9)

Figure 14-9 *My Pet* video.

Wild West Showdown

Stage your own shootout on Main Street, just like you have seen thousands of times in the Wild West movies. Maybe the good guy and bad guy fight over a girl. Maybe the bad guy is just tearing up the town. Maybe the townsfolk won't support the sheriff and he has to take on a whole gang by himself. The scenarios are endless. The results will be fun to do and to watch.

Wish You Were Here

This may be a vacation post card for someone back home or it can be a video note to someone you miss every day.

The Women/Men in My Life

All the men or women in your life. From kids to adults. From relatives to friends to the people you work with. Everyone belongs in this one. Ask them all to say *hi* to the camera and to you.

You Making a Video

You will not be able to shoot all of this one yourself but you can certainly plan it all. It should include *you* taping as well as shots of what you were taping. It could include clips from some of your videos. It should include *you* in different taping postures such as working a handheld camera or a tripod camera, positioning a shiny board, directing talent, preparing props, and editing the finished product. You might want to include some shots just for this video, such as shooting sideways, standing on your head, or from the top of the Golden Gate Bridge. You will, of course, do anything to get the shot. Right? Right.

Zoos and Zebras

Zebras are fascinating animals. How do you suppose they got those stripes? Anyway, whether you care or not, the zoo is always a good place to shoot a video about animals, both the exotic and the human kind. They are all there in one form or another. Maybe you'd like to use the zebra for your title or close, coloring it to make it a rainbow zebra, beyond the limits of the imagination.

CHAPTER FIFTEEN
SCRIPTS TO SHOOT

The scripts in this chapter are put here to be produced by you. Adapt them to your needs. Change the locations, the props, the cast, or whatever to fit what is available to you.

The scripts are written as show scripts, not shooting scripts, meaning that not all shots are noted in the script. The script gives you the basic shot and it is up to you as the producer/director to plan your own shots beyond those written into the script. This is the way the professionals do it and that is what we want to make you, a professional. The way you decide to adapt these scripts will be a reflection of your own style and creativity. You may decide to make the Western Shootout a street fight with laser guns. Our Treasure Hunt may become a search of your own backyard. And the Romantic Interlude may become a scene in your very own kitchen.

If you use one of these scripts, you will not need to do certain steps of the planning techniques outlined in this book. You can bypass the outline, the format, and the script. You will want to consider doing these planning steps though:

Chapter 4 The Budget
            5 The Equipment and the Crew
            8 The Storyboard
            9 The Final Pieces
        10 The Taping Plan
        11 The Taping Day
        12 The Edit

Some of these have been started for you already. We have included with the scripts a brief notation of cast, locations, set, and special props. You can use our cast list and just cast the roles. You can use our location list and decide specifically where you will roll tape, and you can use our prop list. Doing this, you have only to prepare the following:

Budget
Equipment list
Location sketches for camera placement and lighting plot
Crew list
Shot list

You will want to spend most of your time on the shot list, making sure that the show will look the way you want it to. Remember if you need the visualization, you can make storyboards of any or all of the script.

Of course, you will want to do your shot log during taping and your edit list prior to going into the edit session.

Have fun with these scripts. That is, after all, what TV is all about.

POPCORN

A 30-Second TV Commercial

Figure 15–1 Popcorn rides again in this TV commercial.

Copyright JM Production Co. '69

CAST: VOICE-OVER ANNOUNCER
         SECOND VOICE OVER

LOCATION: KITCHEN

SET: STOVE

SPECIAL PROPS: PAN
                 POPCORN
                 OIL
                 PRODUCT PACKAGE

SCENARIO: An up-close look at popcorn exploding, exaggerated with sound and slow motion and set to the theme from *The Lone Ranger*.

NOTES: This can be shot with one camera. Check your lens to make sure that it can focus on the close-ups called for. The slow motion will be achieved in postproduction with slow motion edit capability or you can use the frame advance on your VCR to create the effect. Lay the music and sound effects in post but actually record popcorn explosions and, if possible, distort these to sound bigger, more dramatic.

MUSIC: UNDER THROUGHOUT:
THEME FROM "THE LONE RANGER"

OPEN ON CU OF SIDE OF PAN. SILVER
IN COLOR. POPCORN EXPLODES UP
INTO FRAME IN SLOW MOTION.

SOUND: SINGLE KERNEL OF POPCORN
EXPLODING, EXAGGERATED

MUSIC: FIRST BEAT OF MUSIC TO
EXPLOSION

POPCORN FALLS BACK DOWN
OUT OF FRAME.

POPCORN EXPLODES UP INTO FRAME
IN SLOW MOTION.

SOUND: SINGLE KERNEL OF POPCORN
EXPLODING, EXAGGERATED

MUSIC: SECOND BEAT OF MUSIC

POPCORN FALLS BACK DOWN OUT OF
FRAME.

POPCORN EXPLODES UP INTO FRAME
IN SLOW MOTION, ONE EACH FOR
THE NEXT NINE BEATS OF THE MUSIC.

SOUND: POPCORN EXPLODING. NINE
EXPLOSIONS IN RAPID SUCCESSION
EXAGGERATED.

MUSIC: NEXT NINE BEATS OF THE
MUSIC IN RAPID SUCCESSION

POPCORN FALLS BACK DOWN IN SLOW
MOTION.

MUSIC: PAUSE A BEAT

SINGLE POPCORN EXPLODES INTO
FRAME IN SLOW MOTION.

SOUND: SINGLE KERNEL OF POPCORN
EXPLODING, EXAGGERATED

MUSIC: NEXT SINGLE BEAT OF MUSIC

POPCORN FALLS BACK DOWN IN SLOW
MOTION.

POPCORN EXPLODES UP INTO FRAME
IN SLOW MOTION, ONE EACH FOR
THE NEXT NINE BEATS OF THE MUSIC.

SOUND: POPCORN EXPLODING. NINE
EXPLOSIONS IN RAPID SUCCESSION,
EXAGGERATED

MUSIC: NEXT NINE BEATS OF THE
MUSIC IN RAPID SUCCESSION

POPCORN FALLS BACK DOWN IN
EXTREME SLOW MOTION.

MUSIC: PAUSE A BEAT

POPCORN POPS IN RAPID SUCCESSION,
IN REAL TIME. NO SOUND HEARD.

ANNOUNCER
(Delivered in dramatic western style)
From out of the west in a puff of
smoke comes . . .

SOUND: SLOWLY BRING UP THE
SOUND OF THE CORN POPPING

MUSIC: SLOWLY BRING IN THE NEXT
PART OF THE MUSIC.

a warm friend who believes in truth,
justice, and the American way . . .

POPCORN IS POPPING UP INTO THE
FRAME LIKE CRAZY NOW.

SOUND: UP FULL

MUSIC: UP FULL

> Born on a dark night in a field in Iowa, no one is popcorn's equal.

POPPING BEGINS TO SUBSIDE.

SOUND: BEGINS TO FADE

MUSIC: BEGINS TO FADE

POPPING STOPS.

SOUND: STOPS

MUSIC: STOPS

HOLD ON CU SIDE OF PAN AS LAST KERNEL OF CORN FLOATS DOWN IN SLOW MOTION.

> ANOTHER VOICE
>
> What was that anyway?

CUT TO CU OF PRODUCT, POPCORN CAN/JAR.

> ANNOUNCER
>
> (In his Lone Ranger voice)
> That was (names product).

CUT TO CU BOTTOM OF PAN. ONE PIECE OF CORN SITS THERE. IT EXPLODES IN SLOW MOTION.

SOUND: IN SLOW MOTION, LOUD, EXAGGERATED

> ANNOUNCER
>
> Hi, ho, (product's first name,)
> awayyyyyyyyyyyyy!

CUT TO BLACK.

FINDERS KEEPERS

A Treasure Hunt Game Show

Figure 15–2 Search for treasure in *Finders Keepers.*

APPROXIMATE LENGTH: 10 MINUTES

CAST: HOST
      VOICE-OVER ANNOUNCER
      4 CONTESTANTS: 2 TEAMS OF 2 EACH

LOCATION: TREASURE SEARCH SITE.   You choose one. It must have two separate areas that can be searched for treasure.

SET: HOME BASE.   Set up at the Treasure Search Site. Two contestant podiums are positioned next to each other. They are banked by a strip of colored blinking lights shining into the camera.

SPECIAL PROPS:
      STOP WATCH
      WALKIE TALKIE.   One for each contestant. Preferably a walkie talkie that can be worn like a headset rather than carried. This type of headset walkie talkie is activated by voice and leaves the hands free to search for the treasure.
      TREASURE MAPS.   Two, one for each treasure search site. These can be simple, noting only pertinent information that might help the contestant locate the treasure.
      TREASURE CARDS.   20 total, 10 for each treasure hunt. These cards have the name of a prize written on them as well as the number of points attached to that prize.

SCENARIO: Contestants work in teams of two to find hidden treasure. One contestant stays with the host and reads the treasure map. His teammate runs around looking for the treasure cards based on the directions given to him by the contestant with the map. They communicate with each other over the walkie talkies. After the search, the points of the treasure cards are added up and the team finding the most treasure gets to keep everything they have found.

NOTE: This show is designed to be shot at the treasure site. It has no set other than podiums for the contestants and these could be chairs with high backs turned backward. It is shot ideally with two cameras. Both focus on the home base prior to the treasure hunt, but during the treasure hunt, one runs with the searcher and the other stays with the map reader. The two pieces of video are edited together in the edit session.

OPEN WIDE ON HOME BASE. LIGHTS COME UP TO THE BEAT OF THE MUSIC.

MUSIC: THEME: CHOOSE A SONG WITH A PRONOUNCED BEAT

ANNOUNCER

> This is the game show where what you
> find is what you keep . . .

SUPER: TITLE

ALL LIGHTS ARE UP.

> . . . FINDERS KEEPERS . . .

CUT TO A RAPID PAN OF THE AREA TO BE SEARCHED ON THIS SHOW.

> . . . the show that searches the US for
> treasure.

CUT TO WIDE SHOT OF HOME BASE. HOST ENTERS.

TITLE: OUT

> And here is your host (name of host)

MUSIC: UNDER

HOST

> Hi there. We've hidden lots of treasure so let's
> get to it, Johnny.

ANNOUNCER

> Hey, that's me. And these are our contestants
> for today.

CONTESTANTS ENTER TO THEIR PODIUMS AS THEY ARE INTRODUCED.

> On the Blue Team, (names both contestants)

HOST

> Hey, (names contestants) great to have you
> here.

CUT TO 2 SHOT OF BLUE TEAM CONTESTANTS AS THEY RESPOND.

<u>BLUE TEAM CONTESTANTS</u>

(respond)

CUT TO WIDE SHOT HOME BASE.

<u>ANNOUNCER</u>

And on the Red Team (names contestants)

CONTESTANTS ENTER TO PODIUM.

MUSIC: <u>FADES OUT</u>

<u>HOST</u>

Great to have you here, too, (names contestants)

CUT TO TWO SHOT OF RED TEAM CONTESTANTS AS THEY RESPOND.

<u>CONTESTANTS</u>

(Respond)

<u>HOST</u>

All right, we've got two great teams raring to go find that treasure. Where will we be searching today, Johnny?

<u>ANNOUNCER</u>

Right here . . .

VIDEO EFX: CUT TO A CAMERA RUN-THROUGH OF THE SEARCH SITE. PRETAPE THIS WITH THE CAMERA GOING THROUGH THE ENTIRE SEARCH LOCATION, SPEEDED UP BY USING FAST-FORWARD IN THE EDIT.

MUSIC: <u>RUNNING MUSIC FOR THIS PREVIEW OF SEARCH SITE</u>

(names and describes the search site. This description should not last longer than 15 seconds. For example, " . . . in our own backyard, or yours. We're going to search your house, [names host]. That ought to be fun.")

MUSIC: OUT

CUT BACK TO CU OF HOST.

HOST

Sounds like fun to me.

PULL TO INCLUDE ALL CONTESTANTS.

(to contestants)
Can you kind of pick up while you're there?

CUT TO CU OF CONTESTANTS AS THEY RESPOND.

CONTESTANTS

(respond)

CUT BACK TO WS HOME BASE.

ANNOUNCER

Leave the mess. Look for the treasure.

HOST

Right, let's get to the treasure. Before we
started the show, we flipped a coin and Blue
Team, you go first.

BLUE TEAM

(responds)

HOST

You have your walkie talkies? Put them on.

CUT TO CU AS CONTESTANTS PUT ON THEIR WALKIE TALKIES.

As you know, you will be communicating with
each other over these. One of you will be
reading the map, the other will be searching
for the treasure. There are ten treasure
locations. At each one of them is a Treasure
Card like this.

CUT TO CU OF CARD. SHOWS PRIZE AND POINTS ATTACHED TO THAT CARD.

> Every card represents a prize and each is
> worth a certain number of points.

CUT BACK TO HOME-BASE SHOT.

> You will have three minutes to find as many
> as you can. We will then add up the points to
> determine your score. The team scoring the
> most points gets to keep all the treasure they
> find. But, you must be back at the Start when
> the time runs out or you lose all the treasure
> you have found.
> (To contestants)
> Do you all understand?

CUT TO 2S BLUE TEAM.

> #### BLUE TEAM
> (responds)

CUT TO 2S RED TEAM.

> #### RED TEAM
> (responds)

CUT TO CU HOST.

> #### HOST
> Then let's get to it. Blue Team, you're first.
> Which one of you will be reading the map and
> which will be searching?

CUT TO 2S CONTESTANTS AS THEY CHOOSE.

> #### BLUE TEAM
> (responds)

CUT TO WS.

HOST

Okay, (names Searcher) you will look and
(names Map Reader) you will do the directing.
(names Searcher) you go to the start position.

FOLLOW BLUE SEARCHER TO START POSITION. THEN CUT BACK TO CU HOST
AND BLUE MAP READER.

HOST

Here's your map.
(gives map to Blue map reader)
Johnny tell the Blue Team where they will be
searching for treasure.

CUT TO CU OF MAP.

ANNOUNCER

(describe specific area the team will be
searching at the location . . . For example,
"you're going to get the chance to search
[names Host] living room for ten treasures."
This description should be no longer than 10
seconds.)

CUT BACK TO HOST AND CONTESTANT.

HOST

I will put the stop watch where you can see it.
Get ready. Remember you have 3 minutes.

BLUE TEAM MAP READER GETS READY. HOST PLACES STOP WATCH ON MAP
READER'S PODIUM SO HE CAN SEE IT.

HOST

Go.

MUSIC: RUNNING MUSIC UP

BLUE MAP CONTESTANT TAKES OVER DIRECTING THE SEARCH CONTESTANT.
EFFECT DURING THE RACE, BASE PICTURE IS THE SEARCH CONTESTANT
LOOKING FOR THE TREASURE. CUT INTO A BOX IN LOWER-RIGHT CORNER OF

THIS PICTURE IS THE MAP CONTESTANT OR SPLIT SCREEN TO INCLUDE BOTH CONTESTANTS OR CUT FROM ONE CONTESTANT TO THE OTHER AS THE SEARCH IS IN PROGRESS. YOU WILL NEED TWO CAMERAS TO CAPTURE THIS SEARCH SEGMENT: ONE TO RUN WITH THE SEARCHER AND ONE TO STAY WITH THE MAP READER. YOU WILL PUT THESE TWO SEGMENTS TOGETHER IN THE EDIT SESSION.

SUPER: CLOCK IN UPPER-RIGHT CORNER, IF POSSIBLE. (SOME CAMERAS HAVE THIS ABILITY.)

AS TIME RUNS OUT, THE MAP CONTESTANT MUST TELL SEARCH CONTESTANT WHEN TO GO BACK TO THE START TO BE THERE BEFORE TIME RUNS OUT.

SOUND: END BUZZER AT END OF TIME

MUSIC: FADE OUT

CUT TO WIDE SHOT OF HOME BASE. CONTESTANTS ARE HAPPY, OUT OF BREATH.

HOST

You made it. How many treasures did you
find?

PUSH TO CU OF TWO CONTESTANTS AS THEY COUNT THEIR TREASURE CARDS.

BLUE TEAM

(Counts their treasure cards, reading off the
points and the prize for each one)

CUT TO CU HOST.

HOST

(Congratulates them on their treasure)
Now it's the Red Team's turn.

PULL TO INCLUDE THE RED TEAM.

HOST

Are you ready to see what you can do against
these guys?

                                    RED TEAM

(responds)

                                     HOST

Who will be searching?

CUT TO CU OF RED TEAM.

                                    RED TEAM

(responds)

                                     HOST

And that means that (names other
contestant) you will be reading the map?

                                    RED TEAM

(responds)

CUT TO WS.

                                     HOST

Well, let's get to it. (names Red Team
Searcher) you go to the Start position.

FOLLOW THE RED SEARCHER TO START POSITION.

And Johnny, can you tell us where the Red
Team will be searching?

CUT TO A RUN-THROUGH OF THE SEARCH LOCATION.

                                   ANNOUNCER

(another search site near to but not the same
as the one the Blue Team searched. For
example, "Well, the Blue Team took care of
your living room so we're going to send the
Red Team into your bedroom." This
description should be no longer than ten
seconds.)

CUT BACK TO WS.

HOST

> Watch out for those clothes on the floor.
> Here's your treasure map.

HOST GIVES TREASURE MAP TO THE RED TEAM MAP READER

> Be sure to be back at the Start before your
> time runs out. You have 3 minutes. Ready. Go.

PUTS STOP WATCH ON RED TEAM'S PODIUM WHERE RED MAP READER CAN
SEE IT.

MUSIC: RUNNING MUSIC UP

RED MAP CONTESTANT TAKES OVER DIRECTING THE SEARCH CONTESTANT.
EFFECT DURING THE RACE: SAME AS BLUE TEAM TREASURE HUNT.

SUPER: CLOCK IN UPPER RIGHT CORNER, IF POSSIBLE.

AS TIME RUNS OUT, THE MAP CONTESTANT MUST TELL SEARCH CONTESTANT
WHEN TO GO BACK TO THE START TO BE THERE BEFORE TIME RUNS OUT.

SOUND: END BUZZER AT END OF TIME

MUSIC: FADES OUT

CUT TO WIDE SHOT OF HOME BASE. CONTESTANTS ARE HAPPY, OUT OF
BREATH.

HOST

> (comments on treasure hunt) Count your
> cards and let's see who wins today.

CUT TO CU OF CONTESTANTS AS THEY COUNT THEIR CARDS AND READ THE
PRIZES ATTACHED TO EACH.

RED TEAM

> (count their treasure cards, adding up the
> points and telling us the treasure attached to
> every card.)

HOST

(announces the winner) And here's all that treasure you found.

CUT TO SHOTS OF THE PRIZES AS THE DESCRIPTION OF EACH IS GIVEN.

MUSIC: UNDER PRIZE DESCRIPTIONS

ANNOUNCER

(describes prizes, one by one, until all have been described. Descriptions should be 10 to 15 seconds in length, no longer.)

MUSIC: OUT

CUT BACK TO HOME BASE SHOT.

MUSIC: THEME MUSIC FADES IN

HOST

Another great day.

ANNOUNCER

And more great treasure.

HOST

That's it for today. Come back again to the game show where . . .

CONTESTANTS/HOST/ANNOUNCER

what you find is what you keep.

HOST

See you then.

EVERYONE WAVES GOODBYE AS CAMERA PULLS TO WIDE.

BEGIN CREDITS OVER WIDE SHOT. LOWER LIGHTS SLOWLY EVENTUALLY GOING TO BLACK AND ENDING CREDITS OVER BLACK.

<u>ANNOUNCER</u>

This is (announcer's name) and this has been a (name of producer) production.

FADE TO BLACK.

<u>MUSIC: FADE WITH VIDEO</u>

LOVE FROM THE GRAVE

A Vampire Adventure

Figure 15–3   The vampires are loose in *Love from the Grave.*

Copyright JM Production Co. '87

APPROXIMATE LENGTH: 6 Minutes

CAST: TOM—a teenager out for a good time
        LEE—his girlfriend who will do *almost* anything for him
        VAMPIRE
        TEENAGE BOY
        TEENAGE GIRL

LOCATION: CEMETERY
             CAR

SET: GRAVE STONES

SPECIAL PROPS: NONE

SPECIAL WARDROBE:
    VAMPIRE CAPE
    TWO PAIRS OF VAMPIRE TEETH

SPECIAL MAKEUP:
    WHITE PAINT FOR FACE
    BLOOD

SCENARIO: Two teenagers go to a cemetery on a dare to make out. They find more than love.

NOTES: This can be shot with one camera or more. It will require special spot lighting since it takes place at night. Creative use of lights and fog can make the cemetery even scarier than it really is.

OPEN ON CEMETERY, LATE NIGHT, NO MOON, THE WIND IS BLOWING. TOM
AND LEE ARE WALKING THROUGH THE TOMBSTONES. LEE IS CLUTCHING
TOM.

SOUND: ONLY THE SOUND OF THE WIND AND THE SOUNDS OF THE FEET OF
THE TWO TEENAGERS WALKING ON LEAVES ON THE GROUND.

LEE

(whispering) Tom, I don't like this.

TOM

It's just a graveyard.

LEE

That's what I don't like about it.

SHE CLUTCHES HIS ARM EVEN TIGHTER.

TOM

Will you just relax? We said we'd do it here,
didn't we?

LEE

You said we'd do it here.

TOM

So *I* said. What's the big deal?

LEE

I've heard things about this cemetery.

TOM

That's what makes it exciting. Now relax and
give me a little kiss.

HE PULLS HER AROUND AND KISSES HER. PUSH TO CU OF HER FACE. HER
EYES ARE WIDE OPEN, FRIGHTENED. SHE LOOKS AROUND AS IF SHE IS
EXPECTING TO SEE A GHOST OR SOMETHING WORSE.

PULL TO WIDE SHOT AS THEY KISS. WE SEE SOMEONE BEHIND A TOMBSTONE
NEARBY.

CUT TO CU LEE'S FACE. SHE PUSHES TOM AWAY.

> ### LEE
>
> (frightened) I heard something.

> ### TOM
>
> (exasperated with her) You didn't hear anything.

HE TRIES TO KISS HER AGAIN. SHE RESISTS.

> ### LEE
>
> (insistent) I did, too. It came from over there. (points toward tombstone where man was hiding.)

> ### TOM
>
> (lighthearted pleading) Aw, come on, Lee. There is no one here but us and a bunch of dead people.

HE TRIES TO KISS HER AGAIN.

> ### LEE
>
> Tom, stop it. There's someone here.

> ### TOM
>
> Oh, all right. Where is he?

> ### LEE
>
> (pointing) Over there.

> ### TOM
>
> You stay right here. I'll be back in a second.

CUT TO WIDE SHOT AS HE WALKS TOWARD TOMBSTONE. MAN IS NOT THERE ANY MORE. HE IS OVER BEHIND ANOTHER TOMBSTONE BEHIND LEE.

> ### TOM
>
> (mocking her) I'm looking. I'm looking.

CUT TO CU OF LEE'S FACE. SHE IS TERRIFIED.

Come out. Come out wherever you are.

MARY TURNS TO LOOK ALL AROUND HER. A HAND COMES UP BEHIND HER TOUCHING HER SHOULDER. SHE JUMPS.

                        LEE

        (screams)

SHE TURNS. TOM IS STANDING THERE.

                        TOM

        (laughing) You are so funny.

SHE SLUGS HIM.

                        TOM

        Aw, come on, sweetie, let up.

SHE SLUGS HIM AGAIN.

                        LEE

        I'm gettin' out of here, right now.

CUT TO WIDE AS SHE STOMPS OFF. TOM PURSUES HER. THE MAN HIDING BEHIND THE TOMBSTONE COMES OUT AFTER THEY HAVE GONE. WE SEE HIM FROM A DISTANCE. HE HAS ON A TUXEDO WITH A BLACK CAPE. HE WATCHES THEM GO, THEN THROWS THE CAPE AROUND HIS HEAD AND DISAPPEARS.

CUT TO TOM AND LEE WALKING FAST THROUGH THE CEMETERY.

                        LEE

        I don't know why I ever let you talk me into
        this.

                        TOM

        (playfully) Cause you love me.

HE TRIES TO PUT HIS ARM AROUND HER WAIST. SHE PULLS AWAY. HE PERSISTS.

TOM

Aw, come on.

SHE STARTS RUNNING. HE CHASES AFTER HER. THEY RUN THROUGH THE
CEMETERY, CHASING EACH OTHER, LAUGHING.

SHE GETS AHEAD OF HIM AND HIDES BEHIND A TOMBSTONE. HE LOOKS
AROUND FOR HER BUT CANNOT FIND HER.

TOM

(playfully threatening) Where are you? I'm
gonna get you.

SHE JUMPS UP AND RUNS OFF. HE SEES HER AND THE CHASE IS ON AGAIN.

TOM

You can't get away from me.

SHE HIDES AGAIN. HE LOOKS BUT CANNOT FIND HER.

TOM

Where are you? Come on, Lee. Cut it out.

HE GOES OFF WHERE SHE CAN NO LONGER SEE HIM. SHE SMILES TO
HERSELF AT HAVING OUTWITTED HIM. THERE IS A SOUND BEHIND HER.

SOUND: LEAVES RUSTLE

SHE TURNS THINKING IT IS TOM.

LEE

(laughing) Tom.

THE VAMPIRE LOOMS UP, APPEARING OUT OF NOWHERE, HIS CAPE SPREAD
WIDE. HE SMILES AT HER. SHE IS SO FRIGHTENED THAT SHE CANNOT
SCREAM. ALL SHE CAN DO IS LOOK AT HIM, HER EYES WIDE WITH TERROR
AND HER MOUTH OPEN IN A SCREAM THAT WILL NOT COME.

CUT TO TOM STILL LOOKING FOR HER.

> TOM

Lee, this is not funny anymore. Where are
you?

SOUND: LEE SCREAMS

TOM LOOKS FRANTICALLY AROUND AND THEN RUNS TOWARD WHERE HE
THINKS THE SCREAM CAME FROM. SEES NOTHING AND STOPS DEAD IN HIS
TRACKS.

> TOM

Lee, answer me this minute.

SHE JUMPS UP FROM BEHIND A TOMBSTONE, GRABBING AT HIM.

> LEE

Got ya!

HE JUMPS TWENTY FEET, FRIGHTENED OUT OF HIS MIND.

> TOM

(mad) That is not cute, Lee.

TAKES HER HAND OFF HIS ARM. SHE DISSOLVES INTO LAUGHTER AT HIM.

Let's get out of here.

HE GRABS HER ARM AND PULLS HER ALONG. SHE LETS HIM, SMILING
SWEETLY AND STILL LAUGHING BUT SOFTLY NOW. A SMALL SPLOTCH OF
BLOOD IS ON HER NECK. SHE WIPES IT OFF WITH HER FINGER AND THEN
SUCKS ON THAT FINGER AS THEY RUN.

THEY REACH THE CAR. HE OPENS THE PASSENGER DOOR, SHOVES HER IN,
AND SLAMS THE DOOR.

CUT TO INSIDE THE CAR. LEE SITS THERE QUIETLY, A SWEET LITTLE SMILE
ON HER FACE. TOM GETS INTO THE CAR ON THE DRIVER'S SIDE.

> TOM

You shouldn't do that. You almost scared the
life out of me.

> ### LEE
> Isn't that what you wanted, Tommy dear?

HE FUMBLES WITH THE KEYS TRYING TO GET THEM INTO THE IGNITION.

> ### TOM
> Don't "Tom dear" me. This was supposed to be
> fun.

> ### LEE
> I had fun, Tom dear.

> ### TOM
> (looking over at her accusingly) Running
> around over dead people, you call that fun. I'm
> taking you home.

HE TURNS THE KEY TO START THE CAR. SHE REACHES OVER AND PUTS HER HAND ON HIS.

> ### LEE
> Tom dear, I'm hungry.

HE PUTS THE CAR IN GEAR.

> ### TOM
> I'll buy you a burger.

HE TURNS TO LOOK OUT THE BACK SO HE CAN BACK UP.

> ### LEE
> But, I'm hungry for you, Tom.

HE LOOKS AT HER.

> ### TOM
> (shaking his head) Women, I'll never
> understand you.

HE TURNS THE KEY OFF AND MOVES OVER TO HER SIDE OF THE CAR, PUTTING HIS ARMS AROUND HER.

TOM

Is this what you had in mind?

CUT TO CU OF HER FACE.

LEE

Yes, Tom dear. (smiling)

SHE OPENS HER MOUTH, WE SEE HER VAMPIRE TEETH AND SHE BITES DOWN INTO HIS NECK.

TOM

(screams)

TOM PUSHES HER AWAY. SHE GROWLS.

THE BACK DOOR TO THE CAR OPENS. THE VAMPIRE FROM THE CEMETERY GETS IN.

VAMPIRE

Now can we go get some lunch. I'm starved.

TOM STARES IN HORROR.

CUT TO EXTERIOR OF THE CAR. TOM JUMPS OUT AND RUNS OFF INTO THE NIGHT, SCREAMING TO THE TOP OF HIS LUNGS.

VAMPIRE AND LEE

(laughing wickedly)

CARS STARTS UP AND ROARS OFF DOWN THE ROAD.

SOUND: RUSTLING OF LEAVES

A BOY AND A GIRL ARE WALKING THROUGH THE CEMETERY. SHE IS HANGING ON TO HIM.

BOB

Aw, come on, Marilyn. They're only dead
people. What could happen?

PULL TO WIDE AS THEY CONTINUE TO WALK THROUGH THE CEMETERY.

FADE TO BLACK.

SOUND: LONG PIERCING SCREAM FOLLOWED BY A HAUNTING LAUGH

MUSIC: HAUNTING KIND COMES UP LOW

ROLL CREDITS.

ANNOUNCER

> (Vampire voice) Good evening, friends, from
> (names Producer). Won't you drop over for
> dinner sometime? (laughs)

MUSIC: FADES OUT WITH LAST CREDIT

WHO SHOT JACK?

A Murder Mystery

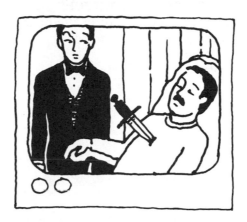

Figure 15–4   Murder in the Reegis House in *Who Shot Jack?*

Copyright JM Production Co. '87

APPROXIMATE LENGTH: 20 Minutes

CAST: JACK REEGIS—handsome beyond belief, rich, young, and notorious as a
ladies' man.
VIRGINIA REEGIS—Mother, loves only herself, wealthy her entire life
and determined to stay that way.
MARY LOU REEGIS—sister and daughter, homely, even her money will
not buy her a husband, hates her brother with a passion.
FRANCESCA SMITH—Jack's girlfriend, tall, gorgeous, blond, in it for the
money.
JONATHON QUADE—the family lawyer, distinguished looking, adores
Virginia or at least adores her money.
MICHAEL SMITH—Francesca's former husband, a bit of a scoundrel,
bitter because Jack has both his wife and money and he has
nothing.
JEEBS—the classic butler in looks, mannerisms, and carriage.
MARIE—the house maid, petite, cute as they come, a bit naive.
GERTRUDE—the Swedish cook, a bit on the overweight side from eating
her own cooking, fun loving, a person everyone loves.
BURT—the chauffeur, tall, lanky, hopelessly in love with Marie.
DETECTIVE SAM JAKES—sloppy appearance, keen mind, dogged
determination to catch the bad guys of the world.
TWO POLICEMEN

LOCATION: THE REEGIS MANSION—A palatial estate, just where you would expect old money to live.

INTERIORS	EXTERIORS
Living room	Front of the house
Library	The woods
Kitchen	
Entryway	

SET: No special set

SPECIAL PROPS:
   A GUN
   A KNIFE
   COMPLETE TEA TRAY

SCENARIO: Jack Reegis is found murdered in the library. Everyone has a motive for knocking him off. Detective Sam Jakes must find the murderer.

NOTES: This script calls for an elaborate mansion. If you do not live in a mansion or have friends that live in one, you can still do this show in a mansion. All you have to do is find one, go there, and tape the exterior at various angles, including a push to the front door. Then you can tape the interiors and even the doorstep somewhere else. Everyone will believe you are in the mansion because that is the exterior you show them. Also, the exterior shots of talent can be done elsewhere by simply shooting these in tight enough that they will cut to the exterior of the mansion.

OPEN ON EXTERIOR SHOT OF THE REEGIS MANSION. IT IS HUGE, OMINOUS, AND COLD. IT IS EARLY EVENING AND THERE IS NO LIGHT COMING FROM ANY OF THE WINDOWS. THE TREES SURROUNDING THE MANSION BLOW VIOLENTLY IN A HIGH WIND AND RAIN THREATENS TO BEGIN AT ANY TIME. SUDDENLY LIGHTNING STREAKS ACROSS THE SKY.

SOUND: INTERMITTENT THUNDER

PUSH TO A WINDOW ON THE UPPER FLOOR.

CUT TO INSIDE THE HOUSE, CONTINUING TO PUSH THROUGH THE ROOM, OUT INTO THE HALL AND DOWN A WINDING STAIRCASE INTO THE LIBRARY. WHEN CAMERA ENTERS LIBRARY, THE CAMERA ABRUPTLY STOPS.

CUT TO DIFFERENT ANGLE TO SHOW JACK REEGIS SITTING IN AN OFFICE CHAIR, THE CHAIR TURNED AWAY FROM THE DOOR. HE SEEMS TO BE STARING OFF INTO SPACE.

SOUND: THUNDER

SNAP ZOOM TO JACK'S FACE. THERE IS NO EXPRESSION ON IT. IT LOOKS DEAD.

AS SOON AS SNAP ZOOM IS COMPLETE, HOLD ONE SECOND THEN,

CUT TO EXTERIOR . . . CAMERA IS IN FIRST PERSON RUNNING THROUGH A WOODED AREA. IT IS ALMOST DARK.

SOUND: SOMEONE CRASHING THROUGH THE UNDERBRUSH

THE PERSON FALLS.

SOUND: PERSON FALLING

CUT TO PERSON ON THE GROUND. THE PERSON IS WEARING A RAIN HAT AND COAT. IT IS IMPOSSIBLE TO SEE WHO IT IS OR TO DISTINGUISH WHETHER IT IS A MAN OR A WOMAN. THE PERSON LOOKS ANXIOUSLY OFF INTO THE DISTANCE.

CUT TO WIDE SHOT OF THE HOUSE. LIGHTNING CUTS A RAGGED PATTERN ACROSS THE SKY.

SOUND: THUNDER

CUT BACK TO THE PERSON IN THE WOODS, HE/SHE GETS UP AND PLUNGES INTO THE BRUSH AGAIN, RUNNING AWAY FROM THE HOUSE LIKE THE DEVIL HIMSELF WERE AFTER WHOEVER IT IS.

CUT TO WIDE SHOT OF THE HOUSE.

FADE TO THE SAME SHOT THE FOLLOWING DAY, EARLY MORNING. THE RAIN HAS STOPPED.

A SPORTS CAR DRIVES UP TO THE HOUSE AND PARKS AT THE FRONT DOOR. FRANCESCA SMITH GETS OUT, TOSSING HER HAIR AS SHE DOES. SHE BOUNDS TO THE FRONT DOOR, OPENS IT, AND GOES IN.

CUT TO MEDIUM CLOSE UP OF BURT, WHO IS POLISHING A CAR. HE OBVIOUSLY WAS ADMIRING FRANCESCA FROM A DISTANCE. HE SMILES TO HIMSELF AND GOES BACK TO HIS POLISHING.

PAN 180 DEGREES TO THE DOOR.

SOUND: WOMAN'S SCREAM

SNAP ZOOM TO THE DOOR.

CUT TO INSIDE: ENTRY. CAMERA PANS TO THE RIGHT, PAUSES, THEN PANS TO THE LEFT (AS IF LOOKING FOR THE LOCATION OF THE SCREAM), STOPPING ABRUPTLY AT THE ENTRANCE TO THE LIBRARY. MOVE TO REVEAL FRANCESCA INSIDE THE LIBRARY, HER HAND TO HER MOUTH, CRYING. SNAP ZOOM TO INSIDE THE ROOM.

CUT TO LIBRARY. FRANCESCA'S BACK IS TO THE CAMERA AND WE ARE SEEING WHAT SHE SEES. JACK STILL SITS EXACTLY THE WAY WE SAW HIM LAST NIGHT. ONLY NOW THERE IS A KNIFE STICKING OUT OF HIS CHEST.

VIRGINIA RUSHES INTO THE ROOM, A NAIL FILE IN HER HANDS. SHE TAKES ONE LOOK AT THE SCENE, SEEMS DISINTERESTED AND BEGINS FILING HER NAILS AGAIN.

VIRGINIA

(showing a total lack of emotion) That takes care of that problem.

FRANCESCA

(seems to be truly upset and horrified) My God, he's dead.

VIRGINIA

Surely looks that way.

JEEBS AND MARIE RUSH INTO THE ROOM. JEEBS STOPS SHORT INSIDE THE DOOR, FORCING MARIE TO STOP.

JEEBS

(To Virginia) Madam, are you all right?

VIRGINIA

(she looks up) Of course. But do call the police, won't you Jeebs? I'm afraid Jack has gone and gotten himself killed. (begins filing her nails again) Really too tawdry for words. It's just like Jack.

SHE CASUALLY LEAVES THE ROOM. JEEBS WATCHES WITH NO EXPRESSION OF EMOTION AT ALL, AS IS BEFITTING A GOOD BUTLER. MARIE RECOILS IN HORROR.

MARIE

(cries softly into her apron) Oh, Mr. Jack . . .

JEEBS LOOKS AT JACK, THEN COLDLY LOOKS ASIDE.

JEEBS

(To Virginia) Yes, madam.

HE GOES TO THE PHONE ON THE DESK, BEING CAREFUL NOT TO GET TOO NEAR JACK AND BEING CAREFUL NOT TO LOOK AT HIM AGAIN.

BURT COMES RUNNING IN AND LOOKS FIRST AT MARIE.

BURT

(concerned for her safety) Are you all right?

MARIE

It's Mr. Jack. (pointing to Jack and sobbing even louder)

JEEBS LOOKS UP FROM THE PHONE AND SHAKES HIS HEAD.

> BURT
>
> Is he?

> JEEBS
>
> Yes, Master Jack has left us.

> BURT
>
> Good.

JEEBS GIVE HIM A SUSPICIOUS LOOK.

> JEEBS
>
> Burt?

> BURT
>
> Well, he deserved it. The louse. After what he
> did to Marie.

> JEEBS
>
> That will be enough of that.

HE MOTIONS TO EVERYONE TO CLEAR THE ROOM.

> JEEBS
>
> Everyone, back to your work.

EVERYONE EXITS EXCEPT FRANCESCA. SHE STILL STANDS EXACTLY WHERE
SHE WAS WHEN WE FIRST SAW HER, STILL STARING DOWN AT JACK.

> JEEBS
>
> (on phone) Would you send someone to Reegis
> House? I'm afraid Mr. Jack has been
> murdered.

WHEN FRANCESCA HEARS THIS SHE LETS GO WITH A TIRADE. THE TEARS
ARE GONE.

FRANCESCA

(angry) The rat. How could he do this to me?
Now what will I do?

SHE STORMS FROM THE ROOM.

JEEBS SMILES TO HIMSELF AS HE WATCHES HER GO.

CUT TO FULL SHOT OF THE ENTRYWAY . . . SOMETIME LATER.

SOUND: KNOCKING

MARIE ENTERS, STILL DABBING AT HER EYES. SHE OPENS THE DOOR. SAM
JAKES IS STANDING THERE.

SAM JAKES

(hesitantly) Aw, I believe you are expecting
me.

HE FLASHES HIS BADGE. MARIE STANDS ASIDE TO LET HIM IN. JAKES
WALKS IN, LOOKS AROUND, AND TURNS BACK TO MARIE.

SAM JAKES

Aw, where is the deceased?

MARIE POINTS TOWARDS THE LIBRARY AND BEGINS CRYING AGAIN. JAKES
NODS, STICKS HIS HEAD BACK OUTSIDE THE DOOR.

SAM JAKES

In here, guys.

TWO POLICEMEN APPEAR IN THE DOORWAY, JAKES POINTS TOWARD THE
LIBRARY AND THE POLICEMEN GO IN THERE.

SAM JAKES

Is your Missus at home?

MARIE IS SOBBING UNCONTROLLABLY BUT MANAGES TO POINT TOWARD THE
STAIRS. VIRGINIA REEGIS IS STANDING THERE WATCHING.

> VIRGINIA

(exasperated) Oh, Marie, stop it.

MARIE PUTS HER APRON UP OVER HER FACE AND RUNS OFF.

> SAM JAKES

And you are?

> VIRGINIA

(haughtily) Virginia Reegis, of course. And
who are you?

SHE BEGINS COMING DOWN THE STAIRS.

> SAM JAKES

Detective Sam Jakes, ma'am, PD. I understand
you have a, uh, problem here.

VIRGINIA STOPS AT THE BOTTOM OF THE STAIRS.

> VIRGINIA

(coldly) Yes, my son is dead.

> SAM JAKES

Ah, yes. You have my deepest sympathies.

> VIRGINIA

That's really not necessary, Detective. He was
a worthless lout.

JAKES DOES NOT KNOW QUITE HOW TO REACT TO THIS COLDNESS.

> SAM JAKES

Ah, yes ma'am. If you will excuse me. I'll just
(motions towards the library) . . . ah.

SHE WAVES HIM ASIDE.

> VIRGINIA

By all means. Get him out of here.

JAKES BACKS TOWARD THE LIBRARY A FEW FEET THEN TURNS AND EXITS
TO LIBRARY. VIRGINIA WATCHES HIM GO.

CUT TO THE KITCHEN. GERTRUDE, MARIE, BURT, AND JEEBS ARE ALL THERE.
GERTRUDE IS PREPARING FOOD. MARIE IS TRYING TO HELP BUT IS
HAMPERED BY HER CONTINUOUS CRYING. JEEBS IS SETTING A SILVER TRAY
FOR TEA. BURT LEANS AGAINST A COUNTER, WATCHING.

> BURT

Things will change now. Now that he's finally
gone.

> MARIE

(cries louder)

> GERTRUDE

Stop that, girl. You'll make yourself sick.

> JEEBS

Yes, stop it now. It is done and cannot be
undone. It is God's will.

> BURT

(laughs) Or at least someone's will. That's for
sure.

> GERTRUDE

Burt, have some respect for the dead.

> BURT

He don't deserve no respect. He didn't do
nothin' to earn it in life and I sure ain't givin'
it to him now he's dead. He deserved what he
got. And you all know it.

> JEEBS

That'll be enough of that kind of talk.

> BURT

I'll talk how I like. He took my Marie, didn't

he? Against her will, weren't it? He got what
was comin' to him.

> JEEBS
>
> You keep talking like that you'll be the one
> they throw in jail for it.

BURT BACKS OFF, HIS HANDS HELD UP AS IF TO STOP SOMETHING.

> BURT
>
> Not me. I didn't do it. Not that I wouldn't but I
> didn't.

> JEEBS
>
> Then you'd best keep your comments to
> yourself or you'll be makin' 'em to a judge.

BURT THROWS UP HIS HANDS AND SULKS OUT.

> MARIE
>
> (between sobs) Mr. Jack weren't no bad man.
> He was wild. But he weren't bad. He couldn't
> help himself. I'm glad someone has helped
> him find his peace but, dear God, you don't
> think it were Burt do you?

> GERTRUDE
>
> No, of course not. That's ridiculous.

GERTRUDE LOOKS AT JEEBS AS IF TO GET HIS APPROVAL. HE LOOKS AWAY,
PICKS UP THE SILVER TRAY AND LEAVES THE KITCHEN.

CUT TO THE LIVING ROOM. VIRGINIA IS SITTING BEFORE A ROARING FIRE,
LOOKING THROUGH A MAGAZINE. JEEBS ENTERS WITH THE TRAY.

> VIRGINIA
>
> Thank you, Jeebs. Will you be kind enough to
> get Mr. Quade on the telephone for me?

> JEEBS
>
> Of course, madam.

HE SITS THE TEA TRAY DOWN ON A TABLE AND EXITS AS MARY LOU
ENTERS. SHE IS WEARING RIDING PANTS AND IS FLUSHED FROM A
HORSEBACK RIDE.

MARY LOU

Mother, who is here?

VIRGINIA

The police, dear. (coldly) Your brother has
had the bad taste to be murdered in the
library.

MARY LOU REACTS LIKE SOMEONE HAS JUST GIVEN HER THE MOST
WONDERFUL PRESENT OF ALL AND THEN TURNS AND RUNS FROM THE
ROOM.

JEEBS ENTERS CARRYING THE TELEPHONE. HE PLUGS IT IN AND HANDS IT
TO VIRGINIA.

JEEBS

(very formal) Your phone call, madam.

VIRGINIA TAKES THE PHONE. JEEBS STANDS BY WAITING.

VIRGINIA

Yes, hello, Jonathon. Get over here. Jack has
been murdered and the police are driving me
crazy.

SHE SLAMS THE PHONE DOWN. JEEBS TAKES IT BACK.

JEEBS

Will you have your tea now, madam?

VIRGINIA

Of course. It is ten, isn't it?

JEEBS

Yes, madam.

JEEBS PUTS THE PHONE DOWN AND BEGINS POURING THE TEA.

CUT TO THE LIBRARY. DETECTIVE JAKES IS LOOKING AT THE BODY.
POLICEMAN #2 IS TAKING PICTURES. POLICEMAN #1 IS DUSTING FOR
FINGERPRINTS. MARY LOU COMES RUNNING IN.

MARY LOU

(happily) Oh my. Oh my. Tick . . . tick . . . tick
. . . Jack. Jack. Jack.

JAKES LOOKS UP AT HER.

SAM JAKES

And you are?

MARY LOU

His sister. Is he? (winks)

SAM JAKES

(not responding to the wink) I'm afraid so.
Poor fellow.

MARY LOU

(delighted) Oh my.

SHE BACKS OUT OF THE ROOM.

SOUND: MARY LOU LAUGHING GLEEFULLY

CUT TO OUTSIDE THE FRONT DOOR. A MAN IS STANDING THERE. HE KNOCKS
ON THE DOOR. THERE IS A PAUSE AND JEEBS ANSWERS.

JEEBS

Sir?

MICHAEL SMITH

I want to see Jack Reegis, right now.
(he winks at Jeebs) Tell him to get out here.

JEEBS

I'm afraid that is impossible, sir.

> MICHAEL

Impossible, is it? Get out of my way.

HE SHOVES HIS WAY PAST JEEBS.

CUT TO INSIDE THE HOUSE. DETECTIVE JAKES WALKS OUT OF THE LIBRARY AS MICHAEL ENTERS.

> MICHAEL

(pointing to the detective) Are you Jack Reegis?

HE DOES NOT GIVE THE DETECTIVE TIME TO RESPOND.

Of course not. You're too old. Where is he, the bum? I'm gonna kill him.

> SAM JAKES

I'm afraid someone has already done that for you.

> MICHAEL

(surprised) What?

> SAM JAKES

Jack Reegis has been murdered.

> MICHAEL

(laughs uproariously) You're kidding me.

> SAM JAKES

I am Detective Sam Jakes of the PD. I do not joke about murder.

> MICHAEL

(delighted with the news) Ha, dead, you say.

> SAM JAKES

Dead as they come.

> MICHAEL

Jack dead. (looking up toward Heaven) There
is a God.

HE TURNS AND LEAVES THROUGH THE FRONT DOOR, LAUGHING OUT LOUD
AS HE GOES.

JEEBS STANDS AT THE DOOR WATCHING ALL THIS. THE DETECTIVE TURNS
TO HIM.

> SAM JAKES

Who was that madman?

> JEEBS

Mr. Smith, sir.

> SAM JAKES

Related to Francesca Smith?

> JEEBS

Her former husband, sir.

THE DETECTIVE NODS, THINKING, AND THEN TURNS AND GOES BACK INTO
THE LIBRARY.

THERE IS ANOTHER KNOCK AT THE DOOR.

SOUND: KNOCK AT THE FRONT DOOR

CUT TO LIVING ROOM. VIRGINIA IS SIPPING HER TEA; MARY LOU IS PACING
BACK AND FORTH.

> MARY LOU

Mother, how can you be so calm? Jack is dead.
Here in this house. It's a disgrace.

> VIRGINIA

It's tea time, dear. Would you like some?

> MARY LOU

Mother, don't be so indifferent. This is going
to create a terrible scandal.

JONATHON QUADE ENTERS THE ROOM AND RUSHES OVER TO GRASP
VIRGINIA'S OUTSTRETCHED HAND.

JONATHON QUADE

Virginia, I am so sorry. The detective told me
about poor Jack.

VIRGINIA

(graciously but coldly) Thank you, Jonathon
dear. It really is too common of him.

JONATHON

What can I do?

VIRGINIA

Get those awful police out of my house and see
that Jack is taken to a proper place
(emphasizes this) away from here.

JONATHON

Yes, of course, right away.

HE GENTLY LETS HER HAND GO AND RUSHES FROM THE ROOM.

MARY LOU

Really, mother. You are too much. (laughs)

VIRGINIA

(exasperated) Oh, please, Mary Lou, stop
laughing so loudly. It's so crude of you. At
least wait until the police are gone.

MARY LOU STARES AT HER A MOMENT AND THEN STOMPS FROM THE ROOM.
VIRGINIA GOES BACK TO HER TEA.

DETECTIVE SAM JAKES ENTERS, TIMIDLY.

SAM JAKES

Excuse me, ma'am.

VIRGINIA LOOKS UP.

> VIRGINIA

Come in, Detective. Are you finished?

SAM JAKES WALKS FURTHER INTO THE ROOM.

> SAM JAKES

Well, almost ma'am.

> VIRGINIA

How did my son die?

> SAM JAKES

He was shot once in the head and stabbed
once in the heart.

VIRGINIA TURNS TO HIM AS IF SHE WERE SURPRISED.

> VIRGINIA

He was stabbed?

> SAM JAKES

Yes, ma'am. And shot.

VIRGINIA BEGINS LAUGHING UNCONTROLLABLY.

> SAM JAKES

Something funny, ma'am?

> VIRGINIA

Of course not, Detective. (she forces herself to
stop laughing) My son is dead.

> SAM JAKES

Yes, ma'am.

HE BACKS OFF AND TURNS TO LEAVE, STOPPING ONCE TO GLANCE BACK AT
VIRGINIA, WHO IS BEGINNING TO LAUGH AGAIN.

CUT TO LIBRARY. THE TWO POLICEMEN ARE STILL THERE. #2 IS LOOKING AT
A WINDOW. #1 IS STARING DOWN LOOKING AT SOMETHING ON THE DESK.

> POLICEMAN #1

Jakes. Take a look at this.

JAKES GOES OVER AND LOOKS DOWN AT THE DESK.

CUT TO CU OF JACK'S HAND. IT HAS A PEN IN A DEATH GRIP.

> SAM JAKES

Did you find anything written down?

> POLICEMAN #1

Not yet.

> SAM JAKES

Well, keep looking. This a funny bunch here.
No one seems to be upset about this murder
except that maid. As a matter of fact, they all
seem glad. He must have been one real nice
guy.

> POLICEMAN #1

As compared to them.

> SAM JAKES

Good point. They're not a real likeable group,
are they?

POLICEMAN #2 MOTIONS TO JAKES.

> POLICEMAN #2

Jakes, take a look at this.

JAKES GOES TO THE WINDOW. THERE IS A TRACE OF MUD ON THE WINDOW
SILL.

> SAM JAKES

So, someone came in, did they? Look outside.
See if you can get a footprint.

POLICEMAN #2 NODS AND EXITS THE ROOM. HE PASSES JONATHON QUADE,
WHO IS ENTERING.

JONATHON

Detective, my name is Jonathon Quade. I am
the family attorney. Is he . . . ?

SAM JAKES

Yes, quite dead.

JONATHON

And here in his mother's house. Can you get
him out of here soon? You see his mother is
most distressed by all this.

SAM JAKES

(sarcastically) Yes, I could see that.

JONATHON

Then you will hurry.

SAM JAKES

We will do our job.

JONATHON

Yes, of course. As soon as you can, please.

HE TURNS AND EXITS.

POLICEMAN #1

Jakes.

JAKES WALKS BACK TO THE DESK. POLICEMAN #1 POINTS TO JACK'S HAND
NOT HOLDING THE PEN. YOU CAN BARELY SEE A BIT OF PAPER CLUTCHED
THERE.

SAM JAKES

He did write us a note. Good boy.

HE PRIES THE HAND OPEN, LOOKS AT THE PAPER, SMILES AND SHOVES THE
PAPER INTO HIS POCKET.

> SAM JAKES

You can wrap this up. Get him out of here. I'll be in the living room.

JAKES EXITS.

CUT TO THE LIVING ROOM. VIRGINIA IS STILL THERE. JEEBS IS TAKING THE TEA TRAY AWAY. JAKES BARGES INTO THE ROOM.

> SAM JAKES

(confronts Jeebs) Ah, you are?

JEEBS PAUSES.

> JEEBS

Jeebs, sir.

> SAM JAKES

Jeebs, will you call all the staff to this room?

JEEBS LOOKS AT VIRIGNIA.

Now, please.

VIRGINIA REEGIS NODS HER HEAD AND JEEBS EXITS.

> VIRGINIA

Really, Detective. Must you order my servants around?

> SAM JAKES

I'm afraid so. Will you call your daughter and anyone else in the house to this room?

> VIRGINIA

Must we?

> SAM JAKES

Yes, we must. Now, please.

VIRGINIA GETS UP FROM HER CHAIR.

VIRGINIA

Very well, but do make it short.

SHE EXITS. JAKES IS LEFT ALONE FOR A MOMENT. HE GOES OVER TO THE TABLE NEXT TO WHERE VIRGINIA HAD BEEN SITTING, OPENS A DRAWER IN THE TABLE, LOOKS INSIDE AND THEN CLOSES THE DRAWER.

HE TURNS TO GAZE AT THE FIRE. AFTER A MOMENT, JEEBS, MARIE, GERTRUDE, AND BURT ENTER. THEY STAND QUIETLY ASIDE. A MOMENT LATER, VIRGINIA, JONATHON, MARY LOU, AND FRANCESCA ENTER. VIRGINIA GOES TO HER CHAIR. JONATHON FOLLOWS HER AND STANDS BEHIND IT. FRANCESCA STANDS OFF TO ONE SIDE, MARY LOU GOES TO THE OTHER SIDE OF THE ROOM. THE SERVANTS CONTINUE TO STAND IN A GROUP.

VIRGINIA

Everyone is here, as you requested, Detective.

JAKES LOOKS AT EACH OF THEM AND THEN PACES BACK AND FORTH AS THEY WATCH. HE PULLS THE PAPER FROM HIS POCKET.

SAM JAKES

Your son left us a note.

VIRGINIA

(gasps aloud)

JAKES LOOKS AT HER. BEFORE HE CAN SAY ANYTHING ELSE, POLICEMAN #1 COMES INTO THE ROOM.

POLICEMAN #1

Jakes.

JAKES WALKS OVER TO HIM. THEY WHISPER TO EACH OTHER. JAKES LOOKS BACK AT THE PEOPLE IN THE ROOM. HE WHISPERS AGAIN TO THE POLICEMAN AND THE POLICEMAN TURNS AND EXITS THE ROOM.

JONATHON

Detective, must we be so dramatic?

JAKES IGNORES HIM.

SAM JAKES

Jack Reegis was murdered last night.

VIRGINIA

(exasperated) For Christ's sake, Detective, we all know that.

SAM JAKES

Twice.

JONATHON

(aghast) Twice?

SAM JAKES

He was shot around 10 last night, I figure, and some time later, after the rain, someone stabbed him.

MARIE

(cries softly)

MARY LOU

Jack died twice. That's appropriate.

SAM JAKES

You didn't like your brother, Miss Reegis?

MARY LOU

(she laughs softly) You could say that. He was a louse.

VIRGINIA

Hush, Mary Lou.

MARY LOU

Well, it's true. And I don't think the Detective believes there was any love lost on Jack in this house anyway.

>   SAM JAKES

I got the feeling that he was not well liked.

>   MARY LOU

Not well liked. You can say that again. He
squandered the family fortune,

VIRGINIA TURNS AWAY. JONATHON CONSOLES HER WITH A PAT ON THE
SHOULDER.

>   he raped the servants . . .

MARIE CRIES LOUDLY AND BURIES HER HEAD IN BURT'S SHOULDER.

>   he beat up on the only woman who would put
>   up with him for any length of time . . .

FRANCESCA THROWS HER HEAD BACK AS IF IT MEANT NOTHING TO HER.

>   even though she never really cared about
>   anything but his money.

>   FRANCESCA

Mary Lou, I'm warning you . . .

>   MARY LOU

Oh, can it Francesca. Everyone knows the
truth. And if that was not enough, he used the
family lawyer to carry on his shady business
deals.

JONATHON STARTS TO SPEAK AND THEN STOPS.

>   MARY LOU

That's right, Jonathon. Keep quiet. No need to
deny it.

>   SAM JAKES

And what did he do to you, Miss Reegis?

>   MARY LOU

Me? (sweetly) Why he was my darling brother

> (hateful) and I hated every bone in his body.
> He drove away or bought off every man that
> ever looked at me. He made me what you see,
> a bitter old woman with no future. My loving
> brother deserved to die.

POLICEMAN #2 ENTERS AND MOTIONS TO THE DETECTIVE.

SAM JAKES

Excuse me a moment.

JAKES EXITS.

CUT TO THE ENTRYWAY.

SAM JAKES

Yes. Did you find it?

POLICEMAN #2

In the woods by the house, just as you
thought.

HE HANDS SOMETHING TO JAKES. JAKES SHOVES IT INTO HIS POCKET.

And Joe is on his way back with the other
guy.

SAM JAKES

Thanks.

HE TURNS TO GO BACK TO THE LIVING ROOM.

CUT TO LIVING ROOM AS BEFORE. JAKES REENTERS. HE WALKS TO THE
FRONT OF THE FIREPLACE.

SAM JAKES

It is clear that any of you could have killed
Jack Reegis. You all had a reason. But we
know who it was.

THE ROOM IS QUIET. ALL LOOK GUILTY. JONATHON STILL STANDS WITH HIS
HANDS ON VIRGINIA'S SHOULDERS. MARIE IS STILL IN BURT'S ARMS.

JONATHON

Detective, please. Who was it?

JAKES PULLS THE PAPER FROM HIS POCKET, OPENS IT UP.

SAM JAKES

Jack managed to write on this paper the
name of the person who shot him.

HE STOPS. VIRGINIA JUMPS UP.

VIRGINIA

What does it say, for God's sake?

SAM JAKES

It says only M, ma'am. He did not live long
enough to finish.

VIRGINIA BREATHES A SIGH OF RELIEF AND SITS DOWN AGAIN.

VIRGINIA

Then you don't know. M could be anyone.
Mary Lou . . .

MARY LOU

Mother, please.

VIRGINIA

Marie . . .

MARIE CRIES LOUDER.

or even Jonathon.

JONATHON PULLS BACK FROM VIRGINIA'S CHAIR.

JONATHON

Virginia, please.

VIRGINIA

Oh, hush, everyone knows your middle name
is Matthew.

SAM JAKES

Or you, ma'am.

VIRGINIA

I beg your pardon.

SAM JAKES

M could be for mother.

VIRGINIA

(laughs) So it could. Good thinking, Detective.

SAM JAKES

Thank you, ma'am.

HE WALKS OVER AND PULLS OPEN THE DRAWER OF THE TABLE.

SAM JAKES

And you might have done it with this.

CUT TO CU OF GUN IN THE DRAWER.

CUT BACK TO WIDE SHOT AS EVERYONE STRAINS TO SEE INSIDE THE
DRAWER.

VIRGINIA

(nonchalantly) I guess I could have.

THE DETECTIVE SHUTS THE DRAWER.

SAM JAKES

Yes, ma'am, you could have but you didn't.

MARY LOU

Then it wasn't Mother?

> SAM JAKES

No.

> VIRGINIA

Of course not. I may have wished him dead but I would never do such a nasty thing.

JAKES PULLS A GUN FROM HIS POCKET. EVERYONE GASPS.

> SAM JAKES

We found this in the woods. We believe it is the murder weapon.

> JONATHON

Then it was someone outside the house.

> SAM JAKES

No, it was someone inside the house.

> FRANCESCA

Who? Get it over with. Who did it?

HE WALKS TOWARD THE SERVANTS.

> SAM JAKES

You did it, Jeebs.

> JEEBS

Sir?

> SAM JAKES

You did it. You shot him.

> JEEBS

But sir, my name does not begin with M.

> SAM JAKES

Perhaps not but you are the manservant in this house, aren't you?

A SMILE SLOWLY CREEPS ACROSS JEEBS' FACE.

CUT TO JEEBS ENTERING THE LIBRARY. HE IS CARRYING A SILVER TRAY WITH AN OBJECT ON IT. JACK IS SITTING AT DESK. HE LOOKS UP.

JACK

Yes, what is it?

JEEBS

Sir, there is a delivery for you.

JACK

(impatiently) Well, just leave it and go.

JEEBS

Yes sir.

JEEBS SITS THE TRAY DOWN ON THE DESK. JACK GLANCES AT IT, GOES BACK TO HIS WORK, THEN LOOKS UP AGAIN QUIZZICALLY. THERE IS A GUN ON THE TRAY. JEEBS PICKS IT UP.

CUT TO THE UNEMOTIONAL FACE OF JEEBS.

CUT TO CU OF GUN AS IT IS RAISED.

CUT TO CU JACK'S FACE AS HE REALIZES WHAT IS TO HAPPEN.

JEEBS FIRES THEN COOLLY PLACES THE GUN BACK ON THE TRAY, WALKS OVER TO THE WINDOW AND UNLATCHES IT, THEN LEAVES THE ROOM, NEVER LOOKING BACK.

CUT BACK TO PRESENT AS POLICEMAN #1 ENTERS WITH MICHAEL SMITH IN TOW.

SAM JAKES

You did it and he (pointing to Michael) came through the window.

CUT TO MICHAEL CLIMBING THROUGH THE WINDOW. HE STABS JACK, PICKS UP THE GUN AND LEAVES AGAIN THROUGH THE WINDOW.

and put the knife in some time later just to

confuse the issue. He took the gun to dispose
of it.

CUT BACK TO PRESENT.

SAM JAKES

But he forgot to take this.

HE HOLDS UP THE SILVER TRAY.

It has gun powder on it. And who else but you,
Jeebs, would carry the gun for a murder on a
silver tray?

MICHAEL STRUGGLES FREE FROM THE POLICEMAN AND RUNS TO THE
OTHER SIDE OF THE ROOM.

MICHAEL

Oh no, not me. It was all him. He paid me. I
wouldn't have done it for free. I hated the man
but I couldn't kill, not for free.

FRANCESCA RUSHES TOWARD MICHAEL AND BEGINS POUNDING HIM ON THE
CHEST.

FRANCESCA

You fool . . . fool. I had it all planned. We could
have had it all.

MICHAEL

(pushing her aside) You think I want his
hand-me-downs?

POLICEMAN #1 RUSHES OVER AND THEY STRUGGLE AS CUFFS ARE PUT ON
MICHAEL.

POLICEMAN #2 ENTERS AND HANDCUFFS JEEBS.

JEEBS

(smiling slyly) You should give me a medal.
The world will be better without him. He was
really worthless, you know.

> ### SAM JAKES
>
> Where did you get the money to pay Michael Smith?

JEEBS SMILES BUT SAYS NOTHING.

JAKES MOTIONS TO THE TWO POLICEMEN.

> Take them away.

AS THEY LEAVE JEEBS LOOKS BACK AT VIRGINIA.

> ### VIRGINIA
>
> Thank you, Jeebs.

> ### JEEBS
>
> Yes, madam.

THE POLICEMEN EXIT WITH MICHAEL AND JEEBS.

> ### MARY LOU
>
> (incredulous) The butler did it. That's hot. Jack would love it. The butler did it.

SHE EXITS LAUGHING TO HERSELF.

JAKES TURNS TO VIRGINIA.

> ### SAM JAKES
>
> Ma'am, I'd unload that gun if I were you. You could kill someone.

> ### VIRGINIA
>
> Thank you, Detective. I don't believe I need it loaded any more anyway. Someone else has taken care of my problem.

JAKES GIVES HER A KNOWING LOOK.

> ### SAM JAKES
>
> Yes, ma'am. I may want to talk to you again.

<u>VIRGINIA</u>

I'll be here.

<u>SAM JAKES</u>

Yes, ma'am.

TURNS AWAY AND EXITS SHAKING HIS HEAD, TALKING TO HIMSELF.

> The butler did it. Never in my career. The
> butler. They're never going to believe this one.

CUT TO EXTERIOR OF HOUSE. THE POLICE CARS ARE PULLING AWAY. IT IS GETTING DARK. LIGHTNING IS TRACING ACROSS THE SKY. THERE IS A SLIGHT DRIZZLE.

<u>SOUND: THUNDER BOOM</u>

COMING FROM THE HOUSE, CLEAR AND COLD, IS SOUND OF MARY LOU'S LAUGHTER.

SAM JAKES LOOKS BACK AT THE HOUSE. SHAKES HIS HEAD, THEN TURNS AND GETS INTO THE CAR. IT DRIVES AWAY.

FADE TO BLACK.

ROLL CREDITS OVER BLACK

<u>SOUND: THUNDER AND LIGHTNING CONTINUES CREDITS</u>

A SOAPY AFFAIR

An Episode From An Afternoon Soap

Figure 15–5 Hot passion in the office in *A Soapy Affair.*

Copyright JM Production Co. '87

APPROXIMATE LENGTH: 3 minutes

CAST: BOB—considers himself God's gift to women and tries to take care of
them all.
CATHERINE—A gorgeous blonde.
CYNTHIA—A gorgeous redhead.
DOROTHY—Bob's secretary, bookish looking, wears her hair up and
always dresses in a business suit, but underneath it all she is a
gorgeous brunette.

LOCATION: BOB'S OFFICE—like any other office

SET: NOTHING SPECIAL

SPECIAL PROPS: HAIR PINS

SCENARIO: Bob loves Catherine, and Cynthia, and Dorothy, or at least he tries.

NOTES: This is not an XXX rated video. Rather it is a fun romp, much as you
would see on your favorite afternoon soap.

OPEN ON WS OF BOB'S OFFICE. HE IS TANGLED UP WITH CATHERINE ON A COUCH. THEY ARE PULLING AT EACH OTHERS CLOTHES PASSIONATELY. HIS SHIRT IS HALF OFF AND HER SKIRT IS ALREADY ON THE FLOOR.

MUSIC: VIOLINS PLAYING SWEET ROMANTIC MUSIC

BOB

Catherine. Catherine. (nibbling on her neck)

CATHERINE

Bob. Bob. (ripping at his shirt and causing the last buttons to fly off.)

THE DOOR OPENS.

MUSIC: STOPS ABRUPTLY

CYNTHIA BARGES INTO THE ROOM. STOPS DEAD STILL WHEN SHE SEES THEM. BOB TAKES ONE LOOK AT HER AND PULLS AWAY FROM CATHERINE.

BOB

(a bit embarrassed) Cynthia. I wasn't expecting you.

CATHERINE GRABS FOR HER CLOTHES AND SMILES SHEEPISHLY AS SHE RUSHES BY CYNTHIA AND OUT OF THE OFFICE.

CYNTHIA

Obviously. (To Catherine as she passes) Slut.

BOB

(still trying to get himself together) Nothing happened here, Cynthia. It's not what you think.

CYNTHIA

(coldly) Oh really. If I can't believe my own eyes, what can I believe?

BOB

Believe *me*, Cynthia.

HE MOVES TOWARD HER.

> You can believe me. Miss Holmes tripped and
> I was only catching her.

CYNTHIA

> You've got to be kidding. Did her skirt fall off
> when she (mocking) fell?

BOB PUTS HIS ARMS AROUND CYNTHIA.

MUSIC: COMING UP SLOWLY . . . TORRID LOVE MUSIC

BOB

> Aw, come on, (baby-talking her) you know
> there is no one for me but you.

SHE STRUGGLES TO GET AWAY BUT HE HOLDS HER TIGHT.

CYNTHIA

> You big lout, get away from me.

BOB

> Aw, Cindy, baby.

HE TRIES TO KISS HER.

MUSIC: MUSIC BUILDS AS THEIR PASSION BUILDS

SHE RESISTS BUT HE INSISTS AND FINALLY GETS HIS WAY. SHE BEATS AT HIM TO GET HIM TO STOP BUT HE DOES NOT AND FINALLY SHE GIVES IN AND RETURNS THE KISS. THE KISS GOES ON AND ON AND BOB BACKS HER DOWN AND EVENTUALLY THEY ARE ON THE FLOOR. THE KISS ENDS. THEY ARE BOTH OUT OF BREATH.

MUSIC: SOFTENS A BIT

SHE LOOKS AT HIM AND THEN STARTS LAUGHING.

CYNTHIA

> Bob, you are such a scoundrel.

> ### BOB
> Isn't that why you love me, baby?

THEY KISS AGAIN AND BEGIN ROLLING AROUND ON THE FLOOR PASSIONATELY.

MUSIC: BEGINS TO BUILD AGAIN AS THEIR PASSION BUILDS AGAIN

THE DOOR OPENS.

MUSIC: STOPS ABRUPTLY

DOROTHY IS STANDING IN THE DOORWAY LOOKING AT THEM.

BOB LOOKS UP.

> ### DOROTHY
> (very businesslike) Bob, do you have a minute?

CYNTHIA STRUGGLES OUT FROM UNDER BOB LEAVING HIM ON THE FLOOR LEANING ON HIS ELBOWS.

> ### BOB
> (unruffled) Sure. Be right with you.

DOROTHY DOES NOT MOVE FROM THE DOOR. CYNTHIA CASUALLY GATHERS UP HER THINGS. SHE IS IN NO HURRY NOR IS SHE EMBARRASSED BY THE SITUATION. BOB GETS UP, GIVES HER A PECK ON THE CHEEK.

> ### BOB
> See you later, kid.

AS SHE WALKS BY HIM TO EXIT, HE PATS HER ON THE BEHIND.

SHE PUSHES BY DOROTHY WHO DOES NOT MOVE.

DOROTHY SHUTS THE DOOR.

SHE STANDS LOOKING INTENTLY AT BOB FOR A MOMENT. HE IS LEANING AGAINST HIS DESK WITH A SLY GRIN ON HIS FACE.

MUSIC: STRIPPER MUSIC

DOROTHY PUTS HER HAND UP TO HER HAIR AND STARTS PULLING OUT HAIR PINS, THROWING THEM AS SHE DOES. HER HAIR FALLS. SHE UNBUTTONS HER BLOUSE SLOWLY AND STARTS WALKING TOWARD BOB. HE LEANS BACKWARD OVER HIS DESK. SHE FOLLOWS HIM DOWN, PINNING HIM TO THE DESK.

                    DOROTHY

        You are such a bad boy.

                      BOB

        And you are late.

                    DOROTHY

        And you are behind schedule.

                      BOB

        We'll just have to make up for lost time.

HE PUTS HIS ARMS AROUND HER AND THEY ROLL OFF THE DESK.

THE DOOR OPENS.

MUSIC: STOPS ABRUPTLY

CUT TO CU OF BOB'S FACE AS HE LOOKS UP, DISHEVELED BUT HAPPY, AND "GETTING INTO" THE CONSTANT INTERRUPTIONS.

                      BOB

        Next.

FADE TO BLACK AS BOB PASSES OUT WITH A SMILE ON HIS FACE.

MUSIC: UP UNDER CREDITS

ROLL CREDITS OVER BLACK.

                    ANNOUNCER

        Stay tuned for more soap from (Producer's
        name)

MUSIC: FADES OUT AS CREDITS END

THE MAZE CRAZE

A Game Show

Figure 15–6   Contestants disappear in *The Maze Craze.*

Copyright JM Production Co. '87

APPROXIMATE LENGTH: 6 minutes

CAST: HOST
       HOSTESS
       SCOREKEEPER
       THREE CONTESTANTS

LOCATIONS: STUDIO

SPECIAL SET REQUIREMENTS: THREE PODIUMS FOR THE CONTESTANTS: can be chairs turned backwards.
     GAME BOARD: 5 feet by 5 feet, consisting of 25 squares, each one 1 foot by 1 foot, positioned 5 across and 5 down, making a huge square. These could be pieces of white cardboard.

SPECIAL PROPS: CLIPBOARD: for score keeper.

SCENARIO: Scorekeeper has a game board drawn on a piece of paper, as shown here. The scorekeeper designates every square either red or white. Of the 25 squares, 20 are designated white and 5 red. A prize is attached to every white square. This information is given to no one, only the scorekeeper knows where prizes are located and the color of the squares.

When the game begins, the hostess stands on the center square. Contestants then take turns telling the hostess where to move on the squares. She can move two spaces, to her left, right, forward, or backward or a combination of any of these, but she cannot move diagonally and once a square has been stepped on, it cannot be stepped on again. If a contestant's turn ends on a white square, the prize at that location goes on the contestant's scoreboard. If it ends on a red square, the contestant is "derezzed" which means he disappears. The last contestant remaining wins the game and all the prizes on his scoreboard. The winning contestant also gets the chance to run the maze for a big prize.

For the second game, the Run the Maze segment, the squares are designated 23 white (safe), one green (treasure), and one red (monster). The contestant physically runs the maze by entering the maze square from any outside position around the square, then moving from one square to the next at his own pace trying to find the green square (treasure). Once he has found it the contestant must get out of the maze without stepping on the red (monster) square. If the contestant successfully gets out of the maze with the treasure, he wins everything. If he steps on the red square, thus getting caught by the monster, he is eaten and disappears like the others. He does not win the treasure of the maze but does get to keep all his prizes from the first game.

Squares in the maze are turned over (the other side is a different color) in game #1 when a contestant ends his turn on one and in game #2 when they are stepped on, to show which squares have been used and cannot be stepped on again.

The scorekeeper's clipboard is given to the voice-over announcer before the show begins so that the announcer can call out the colors of the squares and the prizes as they are found.

NOTES: Lighting. Use light as an effect as well as to light the scene. Put colored gels on lights to create color. Frame the action by placing lighting instruments at the corners of the action and pointing them into the cameras. Use spot effects with or without color gels on the game board to highlight the hostess's movement on this board.

Audio.   You will have to mike this according to how many cameras you have on the shoot. If you have one camera, then provide the host with a handheld mike and let him hold it out to contestants as they speak. If you have two cameras, give the host a handheld and put a stationary mike on a stand in front of the contestants. If you have a third mike, you can give the hostess one, preferably wireless since she will be moving around.

Special effects.   A low smoke added to the run-the-maze segment would add a touch of drama to the scene.

OPEN ON SET, LIGHTS FLASHING ON AND OFF TO THE MUSIC.

MUSIC: UPBEAT FAST PACED: YOUR CHOICE

<u>VOICE-OVER ANNOUNCER</u>

Welcome to the game show where everyone is
likely to disappear . . .

SUPER: TITLE

it's The Maze Craze . . .

HOST ENTERS.

And here's your host . . . (name of the Host)

<u>HOST</u>

Hey, everyone out there. Are you ready to
disappear?

<u>AUDIENCE</u>

(off camera) (shouting) YES!!!

<u>HOST</u>

Then let's get our three suckers, I mean
contestants, in here. Kevin, bring 'em in.

<u>VOICE-OVER ANNOUNCER</u>

Right, our fearless trio today are . . . (brings
in the three contestants, by name only)

CONTESTANTS RUN IN TO THEIR DESIGNATED PODIUMS. EACH CONTESTANT
HAS ON A DIFFERENT-COLORED SHIRT TO EASILY DISTINGUISH AMONG
THEM.

Those are your victims. Now make them
disappear, if you can.

<u>HOST</u>

Oh, I can all right. Just stick with me and I'll
show you how. (To contestants) We are glad
you're here even if only one of you will be
staying. (laughs) Before we get started, let's

take a minute to find out what we can about
each of you. (calls each of the contestants one
at a time to tell us something about where
he/she goes to school, works; wants, desires,
whatever)

CUT BETWEEN CONTESTANTS AND HOST DURING THIS EXCHANGE.

Did you get that all down, Kevin?

VOICE-OVER ANNOUNCER

Got it all. And here's the girl that's gonna help
you find it all. (name of Hostess)

HOSTESS ENTERS, GOES TO THE CENTER OF THE SQUARES.

HOSTESS

Hey, I'm ready to run the maze.

HOST

Right into the next world.

VOICE OVER

Yeah. Yeah.

HOST

Hidden in the squares today are some great
prizes and a few pitfalls. You will each have
the chance to tell (name of Hostess) which
way to move on the squares. You may move
her two spaces to her left, right, forward, or
backward or any combination of these. She
cannot move diagonally. If she ends your turn
on a white box, you will add the prize at that
location to your scoreboard. If you win today's
game, all those prizes will be yours. And you'll
get a chance to run the maze for a really big
prize. But . . .

VOICE-OVER ANNOUNCER

Yeah, but . . . but . . .

HOST

If you end your turn on a red square, you will
be derezzed.

VOICE-OVER ANNOUNCER

That means we will make you disappear, right
here, right before your mama's eyes.

HOST

He loves this so much.

VOICE-OVER ANNOUNCER

Yeah. Yeah.

HOST

So, you got the picture.

CONTESTANTS

(respond)

HOST

Then, let's go. (calls first contestant) _____ ,
which way would you like (the Hostess) to
move on the squares?

CONTESTANT #1

(Responds, such as one forward and one to
her right)

HOSTESS MOVES AS INSTRUCTED. VOICE-OVER ANNOUNCER CALLS OUT THE
COLOR OF THE SQUARE. IF IT IS WHITE . . .

VOICE-OVER ANNOUNCER

White. Safe.

HOST

Safe for now and the prize is . . .

CUT TO THE PRIZE, A PICTURE, VIDEOTAPE, OR WHATEVER. WHILE WE ARE

ON A SHOT OF THE PRIZE, THE HOSTESS WILL FLIP THE MAZE SQUARE TO
SHOW THE OPPOSITE SIDE SO THAT CONTESTANTS WILL KNOW THAT SQUARE
CANNOT BE STEPPED ON AGAIN.

VOICE-OVER ANNOUNCER

(Describes the prize)

CUT BACK TO WIDE OF GAME SET.

HOST

(congratulates contestant and moves on to the
next contestant) (contestant's name) You're
next. Tell (hostess' name) where you want
her to move on the squares.

CONTESTANT TELLS HOSTESS. SHE MOVES AS INSTRUCTED.

IF THE SQUARE IS RED . . .

HOST

And the color is . . .

VOICE-OVER ANNOUNCER

RED! RED! RED! And he's out of here.

HOST

Bye (contestant's name).

VIDEO EFX: BEGIN WITH A MEDIUM SHOT OF THE CONTESTANT. SNAP ZOOM
TO A ECU, RACK FOCUS.

SOUND EFX: ZAPPING OUT KIND OF SOUND.

PULL CONTESTANT OFF THE SET, PULL RACK FOCUS BACK INTO FOCUS. THE
CONTESTANT IS GONE. PULL SLOWLY BACK TO INCLUDE ALL CONTESTANTS IN
SHOT.

SHORTEN THE TIME OF THE RACK FOCUS EFFECT IN THE EDIT IF IT IS TOO
LONG.

HOST

He's out of here.

<u>VOICE-OVER ANNOUNCER</u>

Real gone, man.

GO ON TO THE NEXT CONTESTANT.

CONTINUE UNTIL ONLY ONE CONTESTANT REMAINS. THIS CONTESTANT WINS
ALL PRIZES HE HAS ACCUMULATED AND WINS THE RIGHT TO RUN THE
MAZE HIMSELF FOR A BIGGER PRIZE (CASH MONEY PERHAPS).

<u>HOST</u>

All right, (name of contestant) you're the only
one left. It's all up to you now. You're going to
get to run the maze.

CUT TO SHOT OF THE MAZE. PULSE COLORED LIGHT ON IT FOR EFFECT.

Kevin, tell him about it.

<u>VOICE-OVER ANNOUNCER</u>

Within the maze is a treasure box and inside
that treasure box is this . . .

CUT TO SHOT OF PRIZE IN MAZE.

<u>VOICE-OVER ANNOUNCER</u>

(describes prize)

CUT TO WIDE SHOT, HOST/CONTESTANT/MAZE WITH PULSING LIGHTS. AS THE
VOICE OVER CONTINUES, CUT INTO CUS OF THE CONTESTANT AND THE HOST
AS THEY REACT TO THE DIALOGUE.

But, also in the maze is a monster.
(growls)
Yeah. Yeah. And if you land on the one red
square, the monster will eat you up. Ha! Ha!

<u>HOST</u>

(To Voice Over) You're kidding, right?

<u>VOICE-OVER ANNOUNCER</u>

Wrong. I don't kid about monsters.

                              HOST

He's not kidding (name of contestant) so I
guess you better be real careful.

                          CONTESTANT

(promises to be careful)

HOST TAKES CONTESTANT TO THE EDGE OF THE MAZE.

                              HOST

Okay, you can enter the maze anywhere you
want. Once you are in, step on one square at a
time and wait til we tell you what color the
square is. If you step on the green square, you
have found the treasure, but you must get
safely out of the maze in order to win. Are you
ready?

                          CONTESTANT

(Responds)

                              HOST

Okay, be careful of the monster.

CONTESTANT ENTERS THE MAZE, STEPPING ON ONE SQUARE AT A TIME. THE
VOICE-OVER ANNOUNCER CALLS OUT THE COLOR OF THE SQUARES.

MUSIC UNDER MAZE RUN: OMINOUS

IF THE CONTESTANT STEPS ON A WHITE SQUARE, HE IS SAFE AND MOVES ON.

SOUND: FOR WHITE SQUARE—SAFE HAPPY LITTLE SOUND

IF THE CONTESTANT STEPS ON THE GREEN SQUARE, HE HAS FOUND THE
TREASURE.

SOUND: TREASURE SOUND

                              HOST

Congratulations. You have just found the
treasure of the maze. Now be very careful. You

still have to get out safely. Watch where you
step. The monster still lurks.

VOICE-OVER ANNOUNCER

(growls) I'm gonna get you yet.

THE CONTESTANT CONTINUES STEPPING ON SQUARES.

IF THE CONTESTANT STEPS ON THE RED SQUARE, HE IS EATEN BY THE
MONSTER.

SOUND: MONSTER: BIG UGLY SOUND

VOICE-OVER ANNOUNCER

(Growls happily. The monster has his meal)
Yum. Yum. Let me at him.

VIDEO EFX: MAKE THE CONTESTANT DISAPPEAR WITH EFFECTS. START A
SLOW PAN FROM CU OF THE FEET UP AND THEN HALFWAY UP DO A SNAP
PAN UPWARDS, MAKING IT APPEAR THAT THE MONSTER PULLED THE
CONTESTANT INTO THE MAZE.

SOUND EFX: DISAPPEARING SOUND FOLLOWED BY MONSTER MUNCHING.

VOICE-OVER ANNOUNCER

(munches happily) Mmmmmmmmm . . . good.

HOST

Aw, (contestant's name) . . .

HOSTESS

What happened?

HOST

He's/she's left us.

VOICE-OVER ANNOUNCER

MMmmmmmm. Yeah. But he/she brought
some great prizes with him/her.

HOST

Are you keeping those prizes? (To the hostess) He's keeping those prizes.

HOSTESS

Aw, have a heart.

VOICE-OVER ANNOUNCER

I do. Two of them now. Ha! Ha!

HOST

He's pretty awful but don't worry (contestant's name) took all his prizes with him and he will show up for dinner tonight, Mom.

AUDIENCE

Awwwwwwww.

IF CONTESTANT GETS SAFELY OUT OF THE MAZE WITH THE TREASURE:

HOST

Way to go, (contestant's name)

HOSTESS

Yea (contestant's name)

HOST

You get the treasure. What is it, Kevin?

CUT TO PRIZE AS IT IS DESCRIBED.

VOICE-OVER ANNOUNCER

(describes treasure of the maze)

CUT BACK TO HOST/CONTESTANT.

HOST

And you have all those other great prizes. You beat the maze.

SHOW ENDING . . .

MUSIC: UP AND UNDER

<u>HOST</u>

Thanks for tuning in. We'll be back with more.
Check us out. We're on a . . .

<u>HOST/HOSTESS/VOICE-OVER ANNOUNCER/AUDIENCE</u>

(shouting) MAZE CRAZE!!

<u>HOST</u>

See you next time.

PULL TO WIDE SHOT; ALL WAVE TO THE CAMERA.

ROLL CREDITS.

<u>VOICE-OVER ANNOUNCER</u>

This has been a (name of producer)
production. See you next time when we rezz in
one more time.

VIDEO EFFECT: REZZ OUT LAST VIDEO

SOUND: REZZ OUT SOUND COVERS UP AND TAKES OVER MUSIC

MUSIC: OUT TO SOUND EFFECT

SHUCKS, THIS AIN'T FAIR

A Western Shootout

Figure 15–7 An old-fashioned western shootout happens in *Shucks, This Ain't Fair.*

APPROXIMATE LENGTH: 14 minutes

CAST: MARSHAL MATT GOODGUY—A tall, rangy fellow with rugged good looks, principled to a fault, knows his job and does it in spite of everything.
JOHN SLEEZE—The classic hired gun out to make a name for himself.
MARY SUNSHINE—Sweet, cute, so in love with Marshal Matt that she will forgive him anything, even when she does not understand why he does what he does.
JERRY LEE—The town drunk, trying to stop his drinking ways but just cannot seem to get it together; it is just easier to stay drunk than it is to face life as it really is.
BARTENDER—big, brawny, and tough, used to dealing with drunks and men of violence.
MAYOR
BARBER
UNDERTAKER
SLEEZE'S MEN (2)
TWO MEN IN BAR
TOWNSFOLK

LOCATIONS: RED DOG SALOON
STREETS OF SAN RAMON
MARSHAL'S OFFICE

SPECIAL PROPS: 4 SIX-SHOOTERS WITH
                  BELTS AND CAPS
                  2 RIFLES
                  CLOCK, MARSHAL'S OFFICE
                  CLOCK, RED DOG SALOON
                  MARSHAL'S BADGE

SPECIAL WARDROBE: COWBOY HATS
                     WESTERN-CUT
                     SHIRTS
                     SCARVES

SCENARIO: Marshal Goodguy must face down John Sleeze, reported to be one of the fastest guns in the West, and no one will back him up. Mary begs him to walk away but his principles will not allow him to leave. He is the Marshal and must do his job.

NOTES: This script calls for a Western town with horses. If you do not have access to these, adapt it to what you do have. The street where you live, for example, and bicycles. Use cap guns or water guns.

FADE UP ON WIDE SHOT OF THE STREETS OF SAN RAMON. PEOPLE ARE WALKING ACROSS THE STREET GOING ABOUT THEIR BUSINESS. SEVERAL HORSES WITH RIDERS ARE ON THE STREET. THE MARSHAL IS SAUNTERING UP THE STREET.

CAMERA PUSHES TO FULL SHOT OF THE MARSHAL.

MUSIC: INSPIRATIONAL WESTERN THEME UP FULL

SUPER OVER TITLE: SHUCKS, THIS AIN'T FAIR

CAMERA FOLLOWS THE MARSHAL AS HE STOPS NOW AND AGAIN TO TIP HIS HAT TO TOWNSFOLK. THE MARSHAL IS PLEASANT AND THE TOWNSFOLK SEEM TO LIKE HIM.

MARSHAL

> Howdy. (Calls them by name. Tips his hat to
> the ladies and nods to the men)

TOWNSFOLK

Howdy, Marshal Guy. (Smiling)

MARSHAL GOES INTO A BUILDING MARKED "JAIL & MARSHAL'S OFFICE".

CUT TO INSIDE AS THE MARSHAL ENTERS. IT IS A SMALL ROOM WITH ONE DESK, ONE CHAIR IN FRONT AND ONE BEHIND IT. A GUN RACK FULL OF SHOTGUNS AND A RING OF KEYS HANGS ON THE WALL BEHIND THE DESK. A DOOR LEADS OFF INTO ANOTHER ROOM.

THE MARSHAL TAKES OFF HIS GUN BELT AND HANGS IT ON A NAIL BEHIND THE DESK, PICKS UP THE KEYS AND HEADS FOR THE BACK ROOM.

MUSIC: FADES OUT

CUT TO BACK ROOM. THERE IS A LONE CELL. A MAN IS ASLEEP ON A COT. THE MARSHAL ENTERS TO CELL.

MARSHAL

All right, Jerry Lee, time to face the world.

JERRY LEE

Aw, Marshal, do I have to? It's right nice here.

MARSHAL OPENS THE CELL.

> ### MARSHAL
>
> I 'preciate your nice words, Jerry Lee, but this ain't no hotel. Time to move on.

JERRY LEE RELUCTANTLY GETS UP FROM THE BED. GRABS A JACKET RUMPLED UP ON THE FLOOR. HE TRIES TO TUCK IN HIS SHIRT BUT FINALLY GIVES UP, LEAVING ONE SIDE STILL HANGING OUT.

> ### JERRY LEE
>
> Aw . . . all right, I was gettin' kind of thirsty anyway.

> ### MARSHAL
>
> Try a little steak and eggs, Jerry Lee.

JERRY LEE SHUFFLES OUT OF THE CELL.

> ### JERRY LEE
>
> Sure thing. Whatever you say.

HE EXITS THROUGH THE DOOR.

> See you later. (waving to the Marshal as he leaves the room)

SOUND: FRONT-DOOR SLAM

THE MARSHAL CLOSES THE CELL DOOR BUT DOES NOT LOCK IT.

> ### MARSHAL
>
> (to himself) See you later, Jerry Lee.

MARSHAL WALKS BACK TO HIS OFFICE.

CUT TO OFFICE. A SWEET, INNOCENT LOOKING YOUNG WOMAN IS THERE, WAITING PATIENTLY. SHE IS VIRGINAL IN APPEARANCE, EVERYTHING THAT IS GOOD AND SWEET IN LIFE. THE MARSHAL ENTERS, STOPS ABRUPTLY.

> ### MARSHAL
>
> Mary, I didn't hear you come in.

MARY SUNSHINE

You were busy. I didn't want to disturb you.

MARSHAL

(matter of fact) Right. What is it I can do for
you?

MARY MOVES TOWARDS THE MARSHAL UNTIL SHE IS AS CLOSE AS SHE CAN
GET.

MARY SUNSHINE

You can give up this life.

THE MARSHAL TURNS AWAY.

MARSHAL

I can't Mary. You know I can't. This town
needs me.

MARY SUNSHINE

For what? To keep their drunks?

MARSHAL

Now Mary, that's not fair. Jerry Lee is trying
to stop and we've been over this and over this.
I just can't jump up and leave just because
Sleeze is comin' back to town. It wouldn't be
right.

MARY SUNSHINE

Not even for me?

MARSHAL

It's for you that I stay. You wouldn't like me if
I ran like a dirty coward.

MARY SUNSHINE

If you loved me, you'd stop.

<u>MARSHAL</u>

Shucks, Mary. That ain't fair. You know I love
you.

SHE MOVES TOWARD HIM AND CLUTCHES AT HIS ARM.

<u>MARY SUNSHINE</u>

Then quit. Now! Today.

<u>MARSHAL</u>

I can't.

SHE TURNS AWAY FROM HIM.

<u>MARY SUNSHINE</u>

You mean, you won't.

<u>MARSHAL</u>

All right, Mary, have your way. I won't.

<u>MARY</u>

(starts to cry) I can't stand the thought of you
gettin' shot and for what? This town don't
care.

<u>MARSHAL</u>

You know it's more than that, Mary.

<u>MARY SUNSHINE</u>

Not for me. Not any more.

SHE TURNS AND WALKS TOWARDS THE DOOR. BEFORE SHE EXITS, SHE
TURNS ONCE MORE TO FACE HIM, TEARS STREAMING DOWN HER FACE.

<u>MARY SUNSHINE</u>

You're a good man, Matt. Too good to die for a
town that don't care.

<u>MARSHAL</u>

Mary, please. Try to understand.

MARY SUNSHINE

I do understand, Matt. I just can't stand by
and watch. I'm leavin' on the afternoon stage.

MARSHAL

Don't do this, Mary.

MARY SUNSHINE

It's all I can do. Goodbye, Matt.

MUSIC: THEME FADE UP AND UNDER

SHE TURNS AND RUNS FROM THE OFFICE. THE MARSHAL STANDS STILL,
LOOKING AT THE OPEN DOOR FOR A SECOND. THEN HE BOWS HIS HEAD AND
TURNS AWAY FROM THE CAMERA.

CUT TO A WIDE SHOT OF THE STREET. A LONE RIDER IS RIDING HIS HORSE
UP THE STREET TOWARD THE SALOON. MARY, CRYING AND RUNNING
BLINDLY, RUNS IN FRONT OF HIM, CAUSING THE HORSE TO SHY.

SLEEZE

You gonna get yourself killed, lady.

MARY PAUSES ONLY TO GLANCE UP AND THEN SHE RUNS ON. THE CAMERA
STAYS WITH THE MAN. HE RIDES UP TO THE SALOON, DISMOUNTS, AND TIES
HIS HORSE TO THE HITCHING POST. HE WALKS INTO THE SALOON.

CUT TO INSIDE THE SALOON. SLEEZE ENTERS TO BAR. SLAMS HIS HAND
DOWN ON THE BAR.

MUSIC: STOPS ABRUPTLY AS SLEEZE HITS BAR

SLEEZE

Whiskey, barkeep.

THERE IS ONE BARTENDER AT THE BAR. THE DRUNK THAT THE MARSHAL
LET OUT OF JAIL IS HUNCHED OVER A DRINK AT ONE END OF THE BAR AND
TWO MEN ARE SITTING AT A TABLE, TALKING AND DRINKING. THE
BARTENDER LOOKS AT SLEEZE, REACHES FOR A BOTTLE AND A SHOT GLASS.
PUTS THE SHOT GLASS ON THE BAR AND STARTS TO POUR. SLEEZE STOPS
HIM AND TAKES THE BOTTLE.

SLEEZE

I'll take that.

SLEEZE TAKES A SWIG OF LIQUOR FROM THE BOTTLE AND LOOKS UP AT THE BARTENDER.

SLEEZE

Is he here?

BARTENDER

He's here. (short, clipped speech)

SLEEZE

I heard he was leavin'.

BARTENDER

He ain't leavin'.

SLEEZE

Good, saves me the trouble of chasin' him
down.

MAN AT TABLE

You chase him down? (laughing)

SLEEZE TURNS AWAY FROM THE BAR TOWARD THE MAN AT THE TABLE.

SLEEZE

What was that, pardner?

MAN AT TABLE

I ain't your pardner and he ain't runnin' from
the likes of you.

SLEEZE PUTS HIS HAND ON HIS GUN.

SLEEZE

Smile when you say that, friend.

<u>MAN AT TABLE</u>

I ain't your friend and I ain't smilin' neither.

<u>SLEEZE</u>

Then eat lead.

SLEEZE PULLS HIS GUN AND SHOOTS THE MAN THREE TIMES.

SOUND: GUN SHOTS

THE MAN FALLS TO THE FLOOR, DEAD. THE OTHER MAN AT THE TABLE STARTS TO GET UP.

<u>SLEEZE</u>

You want some, friend?

THE MAN PAUSES FOR A MOMENT LOOKING DOWN AT THE DEAD MAN. THEN HE SLOWLY HOLDS HIS HANDS OUT FROM HIS BODY, AWAY FROM HIS GUN.

<u>SECOND MAN AT TABLE</u>

Not me, friend. I don't want nothin'.

SLEEZE PUTS HIS GUN AWAY.

<u>SLEEZE</u>

Good. Bullets cost money and this town ain't
hardly worth what I already spent. You agree,
friend?

<u>SECOND MAN AT TABLE</u>

He weren't nobody to me.

<u>SLEEZE</u>

I'm glad to hear that. Makes things easier.
You wanta go tell HIM I'm here?

<u>SECOND MAN AT TABLE</u>

Yes sir.

SECOND MAN RUNS OUT OF THE SALOON. SLEEZE TURNS BACK TO THE BAR.

SLEEZE

Right friendly town you got here.

BARTENDER

We like it.

SLEEZE

You don't seem too friendly.

BARTENDER

I got a dead body to move.

BARTENDER COMES OUT FROM BEHIND THE COUNTER AND BEGINS PICKING UP THE BODY OF THE MAN SLEEZE SHOT. HE THROWS IT EASILY OVER HIS SHOULDER. SLEEZE CHUCKLES TO HIMSELF AND TURNS BACK TO HIS BOTTLE.

SLEEZE

(chuckles to himself)

CUT TO MARSHAL'S OFFICE. THE MARSHAL IS AT HIS DESK. IN THE ROOM WITH HIM ARE THREE MEN: THE BARBER, THE MAYOR AND THE UNDERTAKER.

MAYOR

You know he's comin' back.

MARSHAL

I know.

BARBER

Well, what you gonna do about it?

UNDERTAKER

Yeah, Marshal. You can't just ignore the situation. He's already killed ten men, not that I don't mind the business but the way he does it, I could be next.

MARSHAL

Look, there ain't no reason to get all riled up.
I'm here, ain't I, and I'll be here when he
comes back.

MAYOR

That ain't what we heard.

BARBER

Yeah, that pretty little school marm is trying
to get you to leave us, I hear tell.

MARSHAL

(emphatically) I told you I ain't leavin' and I
ain't leavin'.

THE DOOR OPENS AND THE MAN FROM THE SALOON COMES RUNNING IN,
OUT OF BREATH.

SECOND MAN AT TABLE

(out of breath)
Marshal, Marshal.

MARSHAL

What is it, C.J.?

SECOND MAN AT TABLE

He's come back.

BARBER

I knew it.

UNDERTAKER

I better go check my stock.

MAYOR

Yeah, I gotta go, too. Make sure the town
keeps indoors.

<u>BARBER</u>

I better go with you. You might need help.

ALL THREE MEN RUN OUT OF THE ROOM LIKE SCARED RABBITS LOOKING FOR A HOLE TO HIDE IN.

<u>SECOND MAN AT TABLE</u>

Marshal, ain't you gonna stop them? You gonna need help. He already killed Joe Michaels. Shot him dead and he never had a chance.

<u>MARSHAL</u>

They wouldn't be no help anyways. How bout you Slim, you got a gun?

THE MAN BACKS TOWARD THE DOOR.

<u>SECOND MAN AT TABLE</u>

I done lost my gun, Marshal. Otherwise, you know I'd be there for ya.

THE MAN TURNS AND RUNS OUT OF THE OFFICE.

<u>MARSHAL</u>

I know you would but it ain't your job. It's mine.

THE MARSHAL TURNS TO THE RACK OF GUNS AND STARTS PUTTING SHELLS IN ONE OF THEM. THERE IS A LOW TAP AT THE DOOR. THE MARSHAL TURNS QUICKLY, POINTING THE GUN TOWARD THE DOOR.

<u>JERRY LEE</u>

Hey . . . hold it there. It's only me.

JERRY LEE STANDS IN THE DOORWAY.

<u>MARSHAL</u>

You almost got yourself kilt, Jerry Lee.

> ### JERRY LEE
>
> You do seem a bit tense, Marshal.

THE MARSHAL TURNS BACK TO THE GUNS.

> ### MARSHAL
>
> Something you want? I'm a bit busy right now.

JERRY LEE WALKS FURTHER INTO THE ROOM.

> ### JERRY LEE
>
> Yeah, he sent me.

THE MARSHAL TURNS TO FACE HIM.

> ### MARSHAL
>
> Well.

> ### JERRY LEE
>
> He's waitin'. Says he'll meet you on the street at noon.

> ### MARSHAL
>
> Right.

> ### JERRY LEE
>
> Unless you want to get out of town before then.

> ### MARSHAL
>
> Not likely.

> ### JERRY LEE
>
> I didn't figure.

THE MARSHAL PUTS THE RIFLE DOWN ON THE DESK. TAKES OUT HIS SIDEARM AND CHECKS IT FOR BULLETS. JERRY LEE DOES NOT MOVE. FINALLY, THE MARSHAL LOOKS UP AT HIM.

                    <u>MARSHAL</u>

Somethin' else, Jerry Lee?

                    <u>JERRY LEE</u>

Yeah, I figured you might need some help.

THE MARSHAL LOOKS AT HIM, SURPRISED AT THE OFFER BUT GLAD FOR IT
AT THE SAME TIME.

                    <u>MARSHAL</u>

That's right nice of you, Jerry Lee. But there
ain't no reason for you to go gettin' yourself
killed on my account.

                    <u>JERRY LEE</u>

I'd consider it an honor, Marshal.

                    <u>MARSHAL</u>

I reckon there ain't another soul in this town
that'd make that offer.

                    <u>JERRY LEE</u>

(proud of himself) That makes me a better
man than any of them, doesn't it?

                    <u>MARSHAL</u>

Or a bigger fool.

                    <u>JERRY LEE</u>

Shucks, Marshal, let me help. I can shoot. I
used to be good at it before the drink took me.

                    <u>MARSHAL</u>

I heard that about you, Jerry Lee.

                    <u>JERRY LEE</u>

(surprised and pleased) You did!

                    <u>MARSHAL</u>

I did. But I ain't gonna let you die in my place.

JERRY LEE

I don't aim to die.

MARSHAL

I don't aim to either. But I can't let you get
into this.

JERRY LEE HANGS HIS HEAD, DEJECTED.

But I do appreciate the offer. There's not
another *man* left in this town.

JERRY LEE RAISES HIS HEAD PROUDLY.

THE MARSHAL LOOKS UP AT THE CLOCK. IT IS NOW TEN TIL NOON.

CUT TO THE SALOON. THE DEAD BODY IS GONE. THE BARTENDER IS BEHIND
THE BAR POLISHING GLASSES. SLEEZE IS THE ONLY MAN IN THE PLACE.

SLEEZE

Your whiskey is lousy, friend.

BARTENDER

Yep.

SLEEZE

You got something against me, friend?

BARTENDER

Nope.

SLEEZE TAKES ANOTHER SWIG OUT OF THE BOTTLE, EMPTYING IT. HE SLAPS
IT DOWN ON THE BAR. THE BARTENDER TAKES ONE LOOK AT IT AND HANDS
HIM ANOTHER.

SLEEZE

This Marshal of yourn, I hear tell he's good
with a gun?

BARTENDER

He does okay.

SLEEZE

I'm better, you know.

BARTENDER

I hear you're pretty good.

SLEEZE

I'm the best. The best there is.

BARTENDER

Maybe.

SLEEZE JUMPS UP, SLAMS THE BOTTLE DOWN ON THE BAR, DRAWS HIS GUN AND FIRES SIX SHOTS INTO THE MIRROR AT THE BACK OF THE BAR. THE BARTENDER QUIETLY WATCHES, LIKE HE HAD EXPECTED SLEEZE TO DO THIS.

SLEEZE

(shouting) I'm the best there is. And your Marshal is a dead man.

THE BARTENDER DOES NOT RESPOND SO SLEEZE REACHES OVER THE BAR AND PULLS HIS FACE DOWN INTO HIS OWN.

I said I'm the best.

THEY STARE EYE TO EYE IN SILENCE FOR A SECOND OR TWO.

BARTENDER

Could be.

SLEEZE LETS HIM GO.

SLEEZE

(laughing wickedly) Could be, nothing. I'm the best there is and when that clock gets to high noon, I'm gonna blow his brains all over the street and show you all.

CUT TO CU OF CLOCK. IT IS IS NOW FIVE MINUTES TO NOON.

SOUND: TICKING OF CLOCK

CUT TO AN EXTERIOR AT THE MARSHAL'S OFFICE. THE MARSHAL COMES OUT
WITH A RIFLE IN ONE HAND. MARY SUNSHINE COMES RUNNING DOWN THE
STREET AND STOPS HIM AT THE DOOR.

MUSIC: FADES UP AND UNDER

MARY SUNSHINE

Matt, you can't do this.

THE MARSHAL LOOKS DOWN AT HER BUT REMAINS STOIC.

MARSHAL

I can't do nothin' else, Mary.

MARY SUNSHINE

But this town don't care. Not a one of them is
gonna lift a hand to help.

JERRY LEE STEPS OUT OF THE OFFICE, A GUN STRAPPED TO HIS SIDE.

JERRY LEE

I will, ma'am.

MARY LOOKS AT HIM AND TURNS BACK TO THE MARSHAL.

MARY SUNSHINE

John, it's not enough. He ain't even held a gun
in ten years. The Lord only knows if he can
fire one.

THE MARSHAL TAKES HER HANDS.

MARSHAL

Now, Mary, you stop this. I'm doin' what I got
to do and this man . . .

JERRY LEE STRAIGHTENS UP WITH PRIDE AT THIS.

. . . this man is going to back me up.

THE MARSHAL HANDS JERRY LEE THE RIFLE.

MARY DISSOLVES INTO TEARS.

MARY SUNSHINE

But Matt . . .

MARSHAL

Now you go home. I'll be there shortly.

HE SLOWLY PUSHES HER AWAY. SHE LEAVES RELUCTANTLY, LOOKING BACK
OVER HER SHOULDER AS SHE GOES.

JERRY LEE COCKS THE RIFLE AND THE TWO OF THEM BEGIN WALKING
DOWN THE STREET, THE MARSHAL IN FRONT, JERRY LEE FOLLOWING A
LITTLE WAYS BACK, LOOKING AT THE ROOF TOPS AND AROUND CORNERS.

CUT TO THE SALOON. SLEEZE IS STILL AT THE BAR. TWO STRANGERS NOW SIT
AT ONE OF THE TABLES. THE CLOCK STRIKES NOON.

SOUND: CLOCK STRIKING TWELVE

SLEEZE TAKES ANOTHER DRINK FROM THE BOTTLE AND SLAMS IT DOWN ON
THE COUNTER. HE BEGINS CHECKING HIS GUN TO MAKE SURE IT'S LOADED,
CLICKING THE BARREL. HE PUTS IT BACK ON HIS SIDE.

SLEEZE

Barkeep, give me your shotgun.

THE BARTENDER LOOKS AT HIM FOR A SECOND, NOT MOVING.

SLEEZE

The shotgun. (shouting) Now!

THE BARTENDER REACHES UNDER THE BAR, PULLS OUT A SHOTGUN, AND
HANDS IT OVER TO SLEEZE.

SLEEZE

Thanks. You're a very agreeable guy.

HE CHECKS THE BARREL TO MAKE SURE IT IS LOADED. HE NODS TOWARD
THE TWO MEN AT THE TABLE. THEY GET UP. SLEEZE TOSSES THE SHOTGUN

TO ONE OF THEM AND THE TWO MEN LEAVE BY THE BACK DOOR. SLEEZE
TURNS TOWARD THE DOOR AS THE CLOCK JUST FINISHES STRIKING THE
NOON HOUR.

CUT TO THE STREET. THE MARSHAL IS WALKING TOWARD THE SALOON,
FOLLOWED CLOSELY BY JERRY LEE. SLEEZE COMES OUT OF THE SALOON AND
HEADS TOWARD THE MARSHAL. THEY COME WITHIN TWENTY FEET OF EACH
OTHER AND STOP, EACH STARING INTENTLY AT THE OTHER.

SLEEZE

> What's the matter, Marshal? Afraid to take me
> on by yourself?

MARSHAL

> I ain't afraid of you or your kind.

HE MOTIONS FOR JERRY LEE TO BACK OFF.

JERRY LEE MOVES FURTHER OFF TO THE SIDE, BUT HE IS STILL KEEPING
HIS EYES OPEN FOR MEN HIDING.

MARSHAL

> You alone, Sleeze?

SLEEZE

> I don't need no help.

JUST THEN ONE OF SLEEZE'S HENCHMEN FROM THE SALOON POPS UP FROM
BEHIND A BUILDING, GUN DRAWN. JERRY LEE FIRES AND BRINGS HIM
DOWN.

MARSHAL

> You alone now, Sleeze?

SLEEZE

> (mad) I goin' get you, Marshal. You're a dead
> man.

MUSIC: STOPS

THEY BEGIN WALKING TOWARD EACH OTHER. EACH WILL TAKE ABOUT FIVE

STEPS. USE CLOSEUPS HERE OF THE HANDS CLASPED ANXIOUSLY OVER THE GUNS AND THE FACES INTENT ON THE FIGHT.

CUT TO MARY SUNSHINE LOOKING OUT OF A DOORWAY.

CUT TO JERRY LEE'S FACE LOOKING AROUND FOR MORE OF SLEEZE'S MEN.

CUT TO ALLEYWAY. SLEEZE'S OTHER MAN STICKS HIS HEAD OUT AND THEN PULLS IT BACK IN.

CUT TO MARY. SHE HAD SEEN THIS. IT IS OBVIOUS TO HER THAT JERRY LEE HAS NOT. SHE TURNS AROUND. SEES A GUN, GRABS IT.

CUT TO THE TWO MEN WALKING TOWARD EACH OTHER.

CUT TO MARY AIMING THE GUN.

CUT TO THE MAN IN THE DOORWAY AIMING HIS GUN AT THE MARSHAL.

CUT TO ACTION ON STREET . . . MARSHAL DRAWS AND FIRES. SLEEZE GOES DOWN, MORTALLY WOUNDED. HIS GUN IS STILL IN ITS HOLSTER.

AT THE SAME TIME, MARY FIRES AND HITS SLEEZE'S MAN, BRINGING HIM DOWN.

JERRY LEE TURNS TOWARD THE GUNSHOT AND ALMOST SHOOTS MARY BUT HE STOPS HIMSELF.

MUSIC: FADE IN

THE MARSHAL WALKS TOWARD SLEEZE. MARY RUSHES OUT TO HIM.

MARSHAL BENDS DOWN TO SLEEZE.

SLEEZE

You got me, Marshal. I was the best. Now
you're the best . . . til someone gets *you*.

SLEEZE BREATHES HIS LAST. THE MARSHAL STANDS UP.

MARSHAL

You was never the best, Sleeze, til now.

MARY PUTS HER ARMS AROUND THE MARSHAL'S WAIST. HE LOOKS DOWN AT HER AND AROUND THE STREET. THE TOWNSFOLK ARE BEGINNING TO COME OUT OF HIDING.

THE MARSHAL SLOWLY REACHES UP, TAKES HIS BADGE OFF, LOOKS AROUND HIM AND THEN DELIBERATELY THROWS IT INTO THE DIRT AT HIS FEET.

MARY SMILES UP AT HIM. HE PUTS HIS ARMS AROUND HER SHOULDER AND THEY WALK AWAY.

THE MARSHAL MOTIONS FOR JERRY LEE TO JOIN THEM AND HE DOES.

PUSH TO A CU OF THE BADGE.

MUSIC: UP FULL

HOLD ON STREET SCENE AS TOWNSFOLK MILL ABOUT.

ROLL CREDITS.

VOICE-OVER ANNOUNCER

> Shucks, folks, that's all for now . . . from
> (names Producer) Productions. Til next time,
> wear your gun low and keep your aim high.

FADE TO BLACK.

<u>FOR OLD TIME'S SAKE</u>

A Romantic Interlude

Figure 15–8 A touching interlude leaves the lovers still apart in *For Old Time's Sake.*

Copyright JM Production Co. '87

APPROXIMATE LENGTH: 22 minutes

CAST: RICK ROMAN—A rugged guy who is at home anywhere in the world; owns Roman's Place.
CATHERINE CHANDLER—Roman's long-lost love; a sultry, sexy, exotic beauty.
ROBERT CHANDLER—Catherine's husband, a man with a cause, fighting for right with words, not arms.
FRANKIE—The piano player at Roman's Place.
CONSTABLE MILBURN—a bit shady but a regular guy for the most part.
CONSTABLE'S DRIVER
MARIE—Singer at Roman's Place, sweet on Rick but he does not return the feeling.
HASEM FAHER—a slimy character who will do anything for money.
THE FAT MAN—The local gangster boss.
CAB DRIVER
BAR PATRONS

LOCATIONS: ROMAN'S PLACE—a bar and restaurant. *The* place to go and to be seen.
RICK'S OFFICE
AIRPORT

       CONSTABLE'S CAR
       CAB
       FAT MAN'S BEDROOM

SET: Nothing special.

SCENARIO: During the war, Rick and Catherine lost each other. Each thought the other was dead. Now they are thrown together but Catherine is not free. She is committed to another man and neither she nor Rick will hurt her husband so that they can be together again. They must part again.

NOTES: The principal location of this script is a cafe. If you do not have access to a cafe where you can tape, tape the exterior of a cafe that does work. Then set up your own kitchen, dining room, den, or living room to approximate the cafe location and shoot tight. If you keep the framing tight and the cuts from one scene to another frequent, you can make this work.

OPEN ON A WIDE SHOT OF ROMAN'S PLACE. NO ONE IS THERE EXCEPT FRANKIE AND HE IS PLAYING THE PIANO.

MUSIC: ROMANTIC, POIGNANT—ONE OF THOSE SONGS THAT MAKES YOU THINK ABOUT A LOST LOVE. WE'LL CALL IT "RICK'S SONG" THROUGHOUT THE SCRIPT

CAMERA DOES A 360 OF FRANKIE AS HE PLAYS.

ROMAN ENTERS. GOES TO THE PIANO AND PUTS HIS HAND ON THE KEYS. FRANKIE STOPS PLAYING AND LOOKS UP AT ROMAN.

> ROMAN
>
> I thought I told you not to play that.

> FRANKIE
>
> But Mr. Rick, there's not one here.

> ROMAN
>
> I'm here.

HE TAKES HIS HANDS OFF THE PIANO KEYS.

> FRANKIE
>
> I won't play it again.

ROMAN WALKS AWAY TOWARDS THE BAR AND FRANKIE BEGINS TO PLAY ANOTHER SONG.

MUSIC: CLASSIC ROMANTIC TUNE

MARIE ENTERS. SHE WEARS A LONG SLINKY DRESS, READY FOR HER SINGING GIG THAT EVENING AT THE CLUB.

> MARIE
>
> How's it going, Roman?

ROMAN WALKS BEHIND THE BAR AND POURS TWO DRINKS, GIVING ONE TO MARIE AND KEEPING ONE FOR HIMSELF.

> ROMAN
>
> Not bad, kid. You seen anything new?

> MARIE

Only you, Roman. I see only you.

> ROMAN

Don't talk like that, sugar. You know I don't
like it.

> MARIE

I know. But it's true and you can't change
that.

> ROMAN

Maybe not.

THEY DRINK AS FRANKIE PLAYS.

CUT TO A WIDE SHOT OF A SMALL AIRPORT. A SMALL PLANE LANDS.

MUSIC: FADE MUSIC OUT

CUT TO INTERIOR OF AIRPORT. CATHERINE AND ROBERT CHANDLER ENTER
TERMINAL, WALK STRAIGHT THROUGH WITHOUT STOPPING. HE CARRIES
ONE SUITCASE AND A BRIEFCASE. SHE CARRIES ONLY A HANDBAG.

CUT TO EXTERIOR AS THEY EMERGE AND GET INTO A CAB.

THE CAB SPEEDS AWAY.

CUT TO INSIDE ANOTHER CAR SPEEDING DOWN A DARK ROAD. CONSTABLE
MILBURN IS IN THE PASSENGER'S SEAT, ANOTHER MAN DRIVES.

> CONSTABLE

Hurry up, man. We're going to miss them.
Their plane landed five minutes ago.

THE DRIVER NODS HIS HEAD AND SPEEDS UP. PAN TO SHOOT THROUGH THE
FRONT WINDOW, FROM DRIVER'S PERSPECTIVE.

> CONSTABLE

I don't know what I will do now that he is
here. It is just too much. First, Roman. Now

this. They'll have my job for this. Hurry, man, hurry.

THE CAR SPEEDS ON.

CUT TO SHOT OF CONSTABLE'S CAR AS IT SPEEDS BY. CAB SPEEDS BY IN OPPOSITE DIRECTION. CAMERA FOLLOWS CAB.

CUT TO INTERIOR OF THE CAB. CATHERINE AND ROBERT CHANDLER ARE SITTING, STARING FORWARD. ONLY THE SOUND OF THE CAB RADIO IS HEARD.

CAB RADIO

Pickup at Fourth and Main. #54 10 o'clock. Take it. Fifth and Main. Anyone at Roman's, answer . . . (continues)

CATHERINE LOOKS OVER AT ROBERT THEN TURNS AWAY.

CATHERINE

Do we have to do this? There's still time to turn back.

ROBERT LOOKS OVER AT HER, PUTTING HIS HAND ON HERS.

ROBERT CHANDLER

I wish you hadn't come, Catherine. It is too dangerous for you. You should have waited for me in Paris.

SHE LOOKS AT HIM, PUTTING HER FREE HAND OVER HIS.

CATHERINE

I cannot wait while you are in danger. I would rather be with you. Whatever happens.

ROBERT CHANDLER

Then what will be, will be. We both do what we must.

HE PUTS HIS HAND ON HER FACE AND SHE SMILES.

CUT TO AIRPORT. THE CONSTABLE'S CAR COMES TO A SCREECHING HALT. THE CONSTABLE JUMPS OUT AND RUNS INTO THE TERMINAL. THE DRIVER WAITS WITH THE MOTOR RUNNING. THE CONSTABLE IS GONE ONLY A MOMENT. HE COMES RUNNING OUT AND JUMPS BACK INTO THE CAR.

PUSH TO SHOOT THROUGH THE CONSTABLE'S WINDOW.

CONSTABLE

Go, man. We have missed them. We must catch
them before it is too late.

DRIVER

Where to?

CONSTABLE

Don't be a fool. To Roman's and step on it.

THE CAR SPEEDS AWAY.

CUT TO ROMAN'S PLACE. ROMAN IS LEANING ON THE BAR LISTENING TO MARIE SING WHILE FRANKIE PLAYS.

MUSIC: ROMANTIC SONG

SOUND: POUNDING ON DOOR

ROMAN GOES TO THE DOOR.

CUT TO MCU OF ROMAN AT THE DOOR AS IT OPENS TO REVEAL THE FAT MAN.

THE FAT MAN

Roman, you lock your friends out.

ROMAN

Hello, Fat Man. You want something? We're
not open yet.

THE FAT MAN

I know. I'd like to have a little chat.

> ### ROMAN
>
> Do we have something to talk about?

> ### THE FAT MAN
>
> Perhaps more than you know.

CUT TO WIDE SHOT AS THE FAT MAN GENTLY PUSHES BY ROMAN. ROMAN ALLOWS HIM TO ENTER.

> ### THE FAT MAN
>
> Could we sit?

ROMAN GESTURES TOWARD A TABLE.

> ### ROMAN
>
> Of course. (To Marie) Bring us a drink, Marie?
> (To the Fat Man) You'll have?

> ### THE FAT MAN
>
> Seltzer, my boy. The hard stuff is not good for
> the belly. (patting his stomach)

> ### ROMAN
>
> One seltzer, Marie.

SHE NODS AND GOES TO THE BAR TO GET IT. PUSH TO SHOT OF ROMAN AND THE FAT MAN AT THE TABLE.

> ### THE FAT MAN
>
> Roman, my boy, how is business?

> ### ROMAN
>
> Is that what you want to talk about?

> ### THE FAT MAN
>
> Hardly.

> ### ROMAN
>
> Then let's cut the small talk and get to it.

THE FAT MAN

If you like.

ROMAN

I like. Just what is it you want?

THE FAT MAN

(leans in to talk quietly) A man will come to
see you today. Do not talk to him.

ROMAN

That doesn't seem too sociable.

THE FAT MAN

Roman, this is not to be joked about. This man
is dangerous.

ROMAN

What has he done?

THE FAT MAN

He is being sought by all of Europe. He is a
criminal.

ROMAN

And what is his crime?

THE FAT MAN

That doesn't matter. What does matter is that
the police want him and if you help him, they
will get you, too.

ROMAN

And you have this great love for me.

THE FAT MAN

You are a friend, Roman. I only want to warn
you.

ROMAN

I thank you for your *friendly* interest . . .

ROMAN GETS UP FROM THE CHAIR.

And now, I have a club to open.

THE FAT MAN GETS UP.

THE FAT MAN

Of course. May I stay for a bit?

ROMAN

Do whatever you like.

ROMAN WALKS AWAY.

THE FAT MAN

Thank you, my good man.

THE FAT MAN SITS AGAIN. ROMAN WALKS OVER TO THE BAR. MARIE HAS THE SELTZER IN HER HAND.

MARIE

What does he want?

ROMAN

To warn me about a man.

MARIE

He best warn you about himself.

SHE LEAVES TO TAKE THE SELTZER TO THE FAT MAN.

PULL WIDE.

CUT TO CONSTABLE'S CAR DRIVING FAST ON THE CITY STREETS.

THE CONSTABLE IS ANXIOUS, INTENTLY LOOKING OUT THE WINDOW.

                              DRIVER

Why do we pursue these people? What have
they done?

                            CONSTABLE

She has done nothing. He has done
everything.

                              DRIVER

Has he killed a man?

                            CONSTABLE

Would it were that easy.

                              DRIVER

What then?

                            CONSTABLE

He makes people angry. Why he does it is of
no consequence. He angers the wrong people.

                              DRIVER

And why is he here?

                            CONSTABLE

Again, I do not know nor do I care to know. I
know only that I must get him out of this
town as quickly as possible or lose my own
position. To know this is enough.

THE DRIVER NODS AND THE CAR SPEEDS ON.

CUT TO EXTERIOR, CAR SPEEDING AWAY.

CUT TO EXTERIOR OF ROMAN'S PLACE. A CAB PULLS UP IN FRONT.
CATHERINE AND ROBERT CHANDLER GET OUT. HE PAYS THE DRIVER AND
THE CAB DRIVES AWAY.

                         ROBERT CHANDLER

Be brave, my love.

CATHERINE

I am not afraid for myself; only for our life together.

ROBERT CHANDLER

It will not change. I promise it.

HE TAKES HER HAND AND THEY WALK INTO ROMAN'S.

WHEN THEY ARE OUT OF SIGHT, HASEM FAHER STEPS OUT FROM AN ALLEYWAY AND WALKS TOWARDS THE CLUB. HE STOPS AT THE DOOR, LIGHTS A CIGARETTE, TAKES A FEW PUFFS, AND THEN WALKS INTO THE CLUB.

CUT TO INSIDE ROMAN'S. IT IS FULL OF PATRONS NOW, ALL DRINKING AND LAUGHING. MUCH SMOKE HANGS IN THE AIR. FRANKIE PLAYS AT THE PIANO. THE FAT MAN IS STILL AT THE TABLE. MARIE IS STANDING NEXT TO FRANKIE'S PIANO. ROMAN IS NOWHERE TO BE SEEN.

CATHERINE AND ROBERT ARE HUDDLED TOGETHER AT THE DOOR, WAITING.

HASEM FAHER WALKS IN, PASSES BY THEM, AND GOES OVER TO THE FAT MAN'S TABLE. THE CAMERA FOLLOWS HASEM TO TABLE.

HASEM

He is here.

THE FAT MAN

I see. There will be trouble.

HASEM

Oh good. (he rubs his hands together in anticipation) Where there is trouble, there is money to be made.

THE FAT MAN

You are a worm, Hasem.

HASEM

(smiling, enjoying the remark) And it is a nice thing to be. So little is expected of you.

PAN TO SHOW ROMAN ENTERING THE ROOM FROM A SIDE DOOR. HE GOES UP
TO THE CHANDLERS, THEN LOOKS AROUND AS IF SEARCHING FOR A TABLE
FOR THEM. HE CHANGES HIS MIND AND TAKES THEM BACK THROUGH THE
SIDE DOOR HE CAME IN.

PAN BACK TO HASEM AND THE FAT MAN.

HASEM

Yes, I think this will be a good show.

THE FAT MAN

And perhaps turn a profit, also. (laughs
softly)

HASEM

(laughs softly, wickedly)

CUT TO ROMAN'S OFFICE. THE CHANDLERS STAND CLOSE TOGETHER IN THE
DOORWAY. ROMAN IS ACROSS THE ROOM.

MUSIC: FROM THE CLUB, PLAYING SOFTLY

ROMAN

It is dangerous. They know you are here.

ROBERT CHANDLER

It doesn't matter. What I do is worth the risk.

ROMAN

And what is that?

ROBERT CHANDLER

I have a list of names of all the freedom
fighters in Europe. They would pay anything
to get it.

ROMAN

Where is this list?

ROBERT CHANDLER

Here. (pointing to his head)

ROMAN

Then you are indeed in trouble.

ROBERT CHANDLER

I must get this information away from these people before they force me to tell what I do not wish to tell.

ROMAN

And how do you propose to do that? You came into this city on the only plane that is allowed to land.

ROBERT CHANDLER

You must help me.

ROMAN

And what would I do? I am not a freedom fighter.

ROBERT CHANDLER

You are an American.

ROMAN

That does not make me a freedom fighter.

ROBERT CHANDLER

It makes you a believer in freedom. Someone who knows the difference between what you see here in this city and what is possible if men are free.

ROMAN

You are talking to the wrong man. There is a man outside who may help you, for a price.

ROMAN STARTS TO LEAVE BUT IS STOPPED BY CATHERINE.

CATHERINE

Roman, please, there is no one else.

ROMAN

I hope for your sake that that is not true
because I will have no part in this.

SHE BOWS HER HEAD UNHAPPILY AND HE EXITS THE ROOM. ROBERT
CHANDLER PLACES A HAND ON HER ARM.

CUT TO THE CLUB. ROMAN WALKS UP TO THE FAT MAN'S TABLE, BENDS OVER
AND SAYS SOMETHING TO HIM. THE FAT MAN SMILES BROADLY, RISES
FROM HIS CHAIR AND GOES TOWARD THE ROOM WHERE THE CHANDLERS
ARE. HASEM FOLLOWS HIM. ROMAN WALKS IN THE OPPOSITE DIRECTION
TOWARD THE PIANO.

CUT TO ROMAN'S OFFICE. THE FAT MAN ENTERS, FOLLOWED BY HASEM. THE
CHANDLERS TURN.

THE FAT MAN

Ah, my good friends, I understand you have a
problem.

THEY LOOK FRIGHTENED.

Do not be afraid. Roman sent me to help you.

ROBERT CHANDLER MOVES TOWARD THE FAT MAN.

ROBERT CHANDLER

We must continue our journey. We must leave
this city as quickly as possible.

HASEM

They know you are here already. (chuckles
wickedly)

ROBERT CHANDLER

Is this true?

THE FAT MAN

I am afraid so.

ROBERT CHANDLER

Then it is already too late.

THE FAT MAN

Perhaps not. Perhaps we can still arrange
something. Do you have money?

ROBERT CHANDLER LOOKS AT HIS WIFE. SHE OPENS HER PURSE AND SHOWS
THE FAT MAN A DIAMOND AND RUBY NECKLACE.

CATHERINE

I have this.

THE FAT MAN TAKES THE NECKLACE, TAKES A JEWELER'S LOUPE OUT OF
HIS POCKET, AND GOES TO A NEARBY LAMP TO EXAMINE THE NECKLACE
MORE CLOSELY. HASEM GREEDILY WATCHES, FONDLING THAT PORTION OF
THE NECKLACE DANGLING FROM THE FAT MAN'S HANDS.

HASEM

It is very nice. Do you have more?

CATHERINE

No, it is all there is left.

HASEM

Pity.

ROBERT CHANDLER

Will it be enough?

THE FAT MAN LOOKS ONCE MORE AT THE NECKLACE AND THEN STANDS UP.

THE FAT MAN

It will be enough if it is all there is. Where do
you wish to go?

ROBERT CHANDLER

Istanbul. We have friends that will help us
from there.

THE FAT MAN

It will be done. Stay here. I will talk to Roman.

THE FAT MAN EXITS, FOLLOWED BY HAREM AND THE CAMERA FOLLOWS THEM BOTH.

CUT TO BAR. THE FAT MAN ENTERS, FOLLOWED BY HASEM. JUST THEN THE CONSTABLE COMES RUSHING THROUGH THE FRONT DOOR, FOLLOWED BY HIS DRIVER. THE FAT MAN AND HASEM SEE THEM, WAIT FOR THEM TO PASS, AND THEN HURRY OUT THE FRONT DOOR. THE CONSTABLE GOES OVER TO ROMAN AT THE PIANO.

CONSTABLE

Are they here?

ROMAN

Well, hello, Constable. A little late aren't you?

CONSTABLE

Roman, cut the small talk. Are they here?

ROMAN

Who?

CONSTABLE

(impatiently) The Chandlers? Are they here yet?

ROMAN

Give me a minute and I'll check the reservation book.

CONSTABLE

Roman, stop playing with me. Just tell me if they are here.

ROMAN

There is no one here but those you see around you. Just the same old crowd.

CONSTABLE

I hope you are not hiding them, Roman. It would not be good for you.

ROMAN

And why would I hide people I do not even
know?

CONSTABLE

It is said that you know her. This Mrs.
Chandler. That you know her well.

ROMAN

Well, it is said wrong. I know no Mrs.
Chandler, now or ever. Now would you like a
drink?

THE CONSTABLE RELAXES A BIT.

CONSTABLE

Well, I guess I have time for one, but only one.
I must catch this Chandler or you will be
pouring drinks for a new Constable.

ROMAN

We would not want that.

CUT TO ROMAN'S OFFICE. THE CHANDLERS ARE SITTING ON A COUCH,
WAITING.

CATHERINE

Do you think we can trust that man?

ROBERT CHANDLER

We have no choice.

CATHERINE

But he seems so . . . so . . .

ROBERT CHANDLER

I know, but it is our only chance. We have no
time to wait for another.

CATHERINE

You know that I love you above all else. Don't
you, Robert?

ROBERT CHANDLER

I know, Catherine. Do not worry so much. We
have gotten this far. There is only one more
step to take and we will be free.

CATHERINE

Yes, tomorrow we will be free.

THEY HOLD EACH OTHER.

CUT TO EXTERIOR OF THE FRONT OF ROMAN'S PLACE. THE FAT MAN AND
HASEM ARE STANDING THERE IN THE SHADOWS.

THE FAT MAN

Go to my safe and get me the papers there.

HASEM

You are going to help these people?

THE FAT MAN

Of course not but we must make a show of it.
We do not want Roman mad at us, do we?

HASEM

(laughs) Of course not. It is a nice place of
diversion.

THE FAT MAN

Now go and hurry back.

HASEM HURRIES OFF AND THE FAT MAN GOES BACK INTO THE CLUB.

CUT TO INTERIOR. ROMAN MEETS THE FAT MAN AT THE DOOR.

ROMAN

Get them out of my office.

THE FAT MAN

(smiling) My dear boy, I will just as soon as it is safe. For now, the Constable sits just there (pointing to the Constable) and it would be messy to move them.

ROMAN

Just do it as quickly as you can.

THE FAT MAN

So it will be as you wish it.

HASEM COMES RUNNING INTO THE ROOM, SHOVES SOME PAPERS INTO THE FAT MAN'S HANDS. ALL THIS ACTIVITY HAS ATTRACTED THE ATTENTION OF THE CONSTABLE, WHO BEGINS WATCHING THE FAT MAN INTENTLY. THIS IS JUST AS THE FAT MAN HAD PLANNED. NOW KNOWING THAT HE IS WATCHED, HE HEADS STRAIGHT FOR ROMAN'S OFFICE. THE CONSTABLE FOLLOWS. HASEM WATCHES THIS, LAUGHING TO HIMSELF.

CUT TO ROMAN'S OFFICE. THE FAT MAN ENTERS. THE CHANDLERS STAND UP.

THE FAT MAN

Ah, my dear friends, I have it arranged.

HE SHOWS THEM THE PAPERS BUT DOES NOT GIVE THEM UP.

ROBERT CHANDLER

When do we leave?

THE FAT MAN

At once.

THE CHANDLERS ARE VISIBLY RELIEVED. JUST THEN THE CONSTABLE COMES BARGING IN.

CONSTABLE

And what have we here?

THE FAT MAN IS NOT SURPRISED. THE CHANDLERS STEP BACKWARD AND CLOSER TOGETHER.

THE FAT MAN

Constable, do you know my friends?

CONSTABLE

Yes, I believe that I do.

THE FAT MAN

Too bad. (throws up his hands) I had hoped
. . . but that is life. I will leave you now.

ROBERT CHANDLER STARTS TO STOP HIM FROM LEAVING.

ROBERT CHANDLER

But wait. You have . . .

THE FAT MAN

Yes, I have what?

CHANDLER SEES THAT IT IS USELESS TO ASK FOR THE RETURN OF THE
NECKLACE. THIS HAS OBVIOUSLY BEEN A SETUP.

ROBERT CHANDLER

Nothing.

THE FAT MAN

Yes, nothing it is. Good day, my friends.

THE FAT MAN EXITS.

CONSTABLE

And now, Robert Chandler, you are mine.

ROBERT CHANDLER

So it seems.

CONSTABLE

You will stay here. I will return.

HE GOES TO THE DOOR AND EXITS.

ROBERT CHANDLER

We are lost.

CATHERINE

We are together. It is all that matters.

CUT TO OUTSIDE THE ROOM. THE CONSTABLE IS STATIONING HIS DRIVER AT THE DOOR TO GUARD IT. THE FAT MAN AND HASEM ARE AGAIN AT THEIR TABLE. ROMAN SEES THE CONSTABLE AND WALKS OVER TO HIM.

ROMAN

Are you commandeering my office, Constable?

CONSTABLE

Only for a little while. I think you know why.

ROMAN

I do not know why and I do not wish to know why. Go about your business and let me know when I can again go about mine.

CONSTABLE

I appreciate your cooperation, Roman. It makes everything so much easier. I do not wish to see you die for being foolish.

THE CONSTABLE EXITS THE CLUB. ROMAN WATCHES HIM GO.

CUT TO MUCH LATER. THE CLUB IS NOW CLOSED. NO ONE IS THERE EXCEPT ROMAN AND FRANKIE. THEY ARE AT THE BAR, DRINKING.

FRANKIE

Why don't you talk to her while you can?

ROMAN

There is nothing to say any more.

FRANKIE

You give up too easily.

ROMAN

I don't give up at all. That is my flaw. I wish I
could give up.

FRANKIE NODS HIS HEAD IN AGREEMENT. ROMAN PUTS HIS DRINK DOWN
AND GOES OVER TO HIS OFFICE. THE CONSTABLE'S DRIVER IS STILL THERE.

ROMAN

I have some business I must do.

THE DRIVER STANDS ASIDE.

DRIVER

Don't do anything foolish.

ROMAN ONLY LOOKS AT HIM AND GOES THROUGH THE DOOR.

CUT TO INSIDE THE ROOM. ROBERT CHANDLER IS ASLEEP ON THE COUCH.
CATHERINE IS STANDING BY A WINDOW. SHE TURNS AS ROMAN ENTERS.

THEY LOOK AT EACH OTHER FOR A LONG TIME, SAYING NOTHING. THEN . . .

MUSIC: "RICK'S SONG" COMING FROM THE OTHER ROOM

ROMAN

Catherine, you shouldn't have come here.

CATHERINE

I knew nowhere else to go. There was no one
else.

ROMAN

There is nothing I can do.

CATHERINE

I know. We have been lucky and our luck has
just run out. It is not your fault.

ROMAN TURNS HIS BACK TO HER.

                    ROMAN

Was it only for him that you came here?

                   CATHERINE

Only for him.

ROMAN TURNS BACK TO HER AND WALKS CLOSER.

                    ROMAN

I don't believe you. You still care. You have
always cared.

                   CATHERINE

(turning away) Please, Rick, no.

                    ROMAN

(moves even closer) You do. I know you do.

                   CATHERINE

I cannot. I will not. What is past is past.

                    ROMAN

So I am the past. Why? Why did you leave me?
I looked for you for months.

                   CATHERINE

Did you? I looked for you, too. When I came
back and saw the house bombed out, I thought
you had died but I refused to believe it. I
looked everywhere but I could not find you
and finally I did give up. And then Robert
found me.

                    ROMAN

But we have found each other, again. We can
be together again.

                   CATHERINE

It is not that easy. Robert needs me now.

                                ROMAN

                And you, do you need him?

                                CATHERINE

                I care for him, Rick.

HE LOOKS AT HER INTENTLY.

                                ROMAN

                I love you, Catherine.

                                CATHERINE

                (tears running down her face) Rick, please.

HE PUTS HIS HAND ON THE SIDE OF HER FACE AND SHE LOOKS AT HIM
LOVINGLY, HER HEAD FOLLOWING HIS HAND.

IT LAST A MOMENT AND THEN HIS HAND DROPS.

                                ROMAN

                (all business) Did the Fat Man help?

                                CATHERINE

                He took what valuables we had but he brought
                us nothing.

                                ROMAN

                Where do you wish to go?

                                CATHERINE

                Istanbul.

                                ROMAN

                I will arrange it.

HE STARTS TO LEAVE BUT SHE STOPS HIM BY GRASPING HIS ARM.

                                CATHERINE

                Rick, I will always love you.

HE STOPS FOR ONLY A MOMENT AND THEN PULLS HIMSELF FREE AND LEAVES THE ROOM.

MUSIC: STOPS ABRUPTLY

CUT TO THE FAT MAN'S BEDROOM. THE FAT MAN IS ASLEEP AND SNORING. THE ROOM IS DARK, LIT ONLY BY THE MOON FLOODING THROUGH A WINDOW.

ROMAN ENTERS, WALKS ACROSS THE ROOM AND STANDS OVER THE SLEEPING MAN. HE BENDS DOWN, GRASPING HIM BY HIS SHIRT COLLAR. THE FAT MAN WAKES UP.

THE FAT MAN

(frightened) What? Who is it?

ROMAN

Where's the necklace?

THE FAT MAN

(relieved) Roman, my friend. What brings you here so late?

ROMAN GRASPS THE SHIRT COLLAR TIGHTER.

ROMAN

The necklace.

THE FAT MAN

Of course. It is here.

HE PULLS IT FROM BENEATH HIS PILLOW AND HANDS IT TO ROMAN.

I have kept it close so I could deal for their freedom.

ROMAN TAKES THE NECKLACE AND TURNS TO LEAVE.

Are you going to help them?

ROMAN LEAVES. THE FAT MAN SHRUGS, ROLLS OVER, AND ALMOST INSTANTLY BEGINS SNORING AGAIN.

CUT TO ROMAN'S PLACE, LATER STILL. ROMAN IS SITTING IN A CHAIR IN FRONT OF THE DOOR TO HIS OFFICE. THE CONSTABLE'S DRIVER IS GONE.

THE CONSTABLE COMES RUSHING IN.

CONSTABLE

Roman, what are you doing?

ROMAN

Helping out a friend.

CONSTABLE

It is too dangerous to help these friends.

ROMAN

It is you that I am helping.

CONSTABLE

How do you figure that?

ROMAN

These are dangerous people, are they not?

CONSTABLE

They are.

ROMAN

And dangerous people can make your own life dangerous, is that not so?

CONSTABLE

This is so. Already, my position is in jeopardy, thanks to this man.

ROMAN

Then why do you bother with him? Let someone else face the dangers of having him in custody.

  CONSTABLE

This makes sense, Roman, but it is too late. It
is already known that they are captured and
held here.

                                         ROMAN

Let me help you. I will take the blame for
their escape.

                                        CONSTABLE

I am a reasonable man. I can see that what
you say makes sense but how can I do it?

                                         ROMAN

Just turn your back. I will do the rest. And,
oh, by the way, here's a trinket I was forced to
confiscate. I am sure you will know what to
do with it.

ROMAN HANDS THE NECKLACE TO THE CONSTABLE. HE TAKES IT, TURNS IT
LOVINGLY OVER IN HIS HANDS, HIS EYES WIDE WITH EXCITEMENT.

                                        CONSTABLE

Yes, I begin to see that your plan will work.

                                         ROMAN

Then go have a drink and leave it to me.

THE CONSTABLE GOES OVER TO THE BAR. ROMAN EXITS TO HIS OFFICE.
FRANKIE ENTERS FROM A BACK ROOM.

                                        FRANKIE

Constable.

                                        CONSTABLE

Frankie. Play us a tune, will you? We have
things to celebrate.

FRANKIE GOES TO THE PIANO AND BEGINS TO PLAY.

MUSIC: ROMANTIC TORCH SONG

CUT TO INSIDE ROMAN'S OFFICE. ROBERT CHANDLER STILL SLEEPS. ROMAN HAS JUST ENTERED. CATHERINE WALKS OVER TO HIM.

MUSIC: CAN BE HEARD SOFTLY PLAYING FROM THE OTHER ROOM

                    ROMAN

        It is done.

                    CATHERINE

        When?

                    ROMAN

        Now.

SHE LOOKS AT HIM FOR A MOMENT AND THEN MOVES INTO HIS ARMS. HE HOLDS HER TIGHTLY AND THEY KISS PASSIONATELY.

                    CATHERINE

        Oh, Rick, I don't want to go.

                    ROMAN

        Hush.

HE KISSES HER AGAIN.

                    CATHERINE

        Will I never see you again?

                    ROMAN

        Never.

HE HOLDS HER AS TIGHT AS HE CAN AND SHE SOBS UNCONTROLLABLY. HE ALLOWS THIS FOR A MOMENT AND THEN GENTLY PUSHES HER AWAY.

                    ROMAN

        You must wake him, now.

                    CATHERINE

        How will I leave you? I cannot.

ROMAN

You will because you must. Now wake him.

SHE LOOKS AT HIM ONCE MORE AND THEN GOES OVER AND WAKES ROBERT CHANDLER.

ROBERT CHANDLER

Yes, have they come for us?

ROMAN

I have arranged for you to leave. The police will allow you to go.

ROBERT CHANDLER

But how?

ROMAN

It does not matter how. You must leave now.

ROBERT CHANDLER LOOKS AT HIM, SEES THAT ROMAN MEANS IT. HE GETS UP, PICKS UP THEIR BAG AND HIS BRIEFCASE.

ROBERT CHANDLER

Then now it will be.

THEY ALL EXIT.

CUT TO CAFE AS THE CHANDLERS AND ROMAN ENTER FROM THE OFFICE. THE CONSTABLE IGNORES THEM.

ROMAN

There is a car outside. It will take you where you need to go.

ROBERT CHANDLER

(shakes Rick's hand) Thank you. You have saved many lives this night.

CHANDLER TURNS AND EXITS TO OUTSIDE.

CATHERINE STANDS LOOKING ONE LAST TIME AT ROMAN.

CATHERINE

Rick, I . . .

ROMAN

It is over. It died in an air-raid long ago. Only
the memories live. And I have many of those.
Go.

CATHERINE

I will not give up. I will see you again.

SHE TURNS TO GO AND THEN TURNS BACK.

CATHERINE

I will always love you, Rick.

TEARS BEGIN TO FALL DOWN HER FACE AND SHE TURNS TO LOOK AT HIM
ONE LAST TIME. THEN SHE EXITS.

ROMAN STANDS LOOKING AT THE CLOSED DOOR FOR A MOMENT.

CONSTABLE

Roman, have a drink.

ROMAN SLOWLY TURNS AWAY FROM THE DOOR AND WALKS TOWARDS THE
BAR.

ROMAN

Yes.

HE SITS DOWN AT THE BAR NEXT TO THE CONSTABLE, TAKES THE OFFERED
DRINK, AND TURNS TO FRANKIE.

ROMAN

Play it for me, Frankie. For old time's sake.
Play it just once.

FRANKIE

Yes, Mr. Rick.

MUSIC: UP FULL "RICK'S SONG"

ROMAN AND THE CONSTABLE SIT QUIETLY AT THE BAR LISTENING.

ROLL CREDITS

VOICE-OVER ANNOUNCER

This has been a (Producer's name)
production, sweetheart.

APPENDIX

Show Title _____

Date _____

MASTER CHECKLIST

(Add or delete items as appropriate to your show)

(Check when completed)   (Date completed)

_____ OUTLINE   _____

_____ BUDGET   _____

_____ EQUIPMENT LIST   _____

_____ CREW LIST   _____

_____ SCENARIO   _____

_____ FORMAT   _____

_____ SCRIPT   _____

_____ STORYBOARD   _____

_____ CAST LIST   _____

_____ MAKEUP   _____

_____ WARDROBE   _____

_____ PROP LIST   _____

_____ LOCATION LIST   _____

_____ SET DESIGN   _____

_____ SHOT LIST   _____

_____ SCHEDULE   _____

_____ LOCATION SKETCHES   _____

_____ CAMERA POSITIONS   _____

_____ LIGHTING PLOT   _____

_____ BIBLE   _____

_____ SHOT LOG   _____

_____ CREDITS   _____

_____ EDIT LIST   _____

_____ DUBS   _____

Figure A–1 Master Checklist.

SHOW OUTLINE

DATE: _____

SUBJECT: _____

PURPOSE: To document _____   To entertain _____

            To inform _____   To motivate _____

       To make the audience happy _____   Sad _____   Think _____

ROLES: _____   TALENT _____

           _____   TALENT _____

           _____   TALENT _____

           _____   TALENT _____

           _____   TALENT _____

           _____   TALENT _____

THE STORY

THE PLOT: _____

_____

      SUBPLOT: _____

THE CLIMAX: _____

_____

      SUBCLIMAX: _____

THE CLOSE: _____

CREDITS: yes _____   no _____

      Over close _____   black _____   show scenes _____

THE OPEN: over black _____   story _____   action sequence _____

      What does it look like? _____

THE STYLE

THE PICTURES: _____

_____

Figure A–2   Show Outline.

THE MUSIC: yes _____ no _____ name it _____

instrumental _____ lyrics _____

THE SOUND EFFECTS: yes _____ no _____ what? _____

SPECIAL ELEMENTS: slides _____ snapshots _____ film _____

video _____ graphics/artwork _____ special effects _____

what? _____

THE LOCATIONS: _____ DATE _____

_____ DATE _____

_____ DATE _____

_____ DATE _____

_____ DATE _____

_____ DATE _____

THE LENGTH: less than 3 min. _____ 3 to 5 min _____

5 to 10 min _____ 10 to 15 min _____ 15 to 20 min _____

over 20 min (specify) _____

SCREENING DATE: _____

IF THIS IS A FAMILY OR FRIENDLY ORIENTED VIDEO, ASK YOUR PRINCIPAL
TALENT . . .

WHAT IS YOUR FAVORITE:

Color _____

Song _____

Actor/actress _____

TV show _____

Movie (classic) _____

Place _____

Other _____

WHAT VISUALS OF THE SUBJECT CAN YOU PROVIDE?

Snapshots Yes _____ No _____ What? _____

Film Yes _____ No _____ What? _____

Video Yes _____ No _____ What? _____

Art Yes _____ No _____ What? _____

Figure A–2 (*cont.*)

Show Title _____

Date _____

BUDGET WORKSHEET

Note: Estimate all costs. It is not necessary to make any final decisions about locations, equipment, crew, edit, or set at this time. This budget is a best-guess estimate of all expenses for the proposed video.

VIDEOTAPE

    Cost per tape $ _____

    No. Cassettes   × _____

                       Total $ _____

LOCATION FEE

(Place, if you know)

_____ $ _____

_____   _____

_____

                       Total $ _____

TRANSPORTATION

    Gas $ _____

    Parking   _____

    Toll charges   _____

    Airline tickets   _____

    Misc.   _____

                       Total $ _____

PROPS

    Buy $ _____

    Rent $ _____

                       Total $ _____

Figure A–3   Budget Worksheet.

EQUIPMENT

(Name it, if possible)

Video _____

_____

_____ $ _____

Audio _____

_____

_____ _____

Lighting _____

_____

_____ _____

Misc. _____

_____ _____

Total $ _____

CREW

Cost per $ _____

No. crew members × _____

Subtotal _____

Plus other + _____

Total $ _____

EDIT

Rent/buy equip _____

_____ $ _____

Subtotal $ _____

Postproduction facility

cost per hour $ _____

No. hours est. × _____

Subtotal $ _____

Total $ _____

Figure A–3 *(cont.)*

TELEPHONE

    Estimated cost $ _____ $ _____

ARTWORK/GRAPHICS

    Misc. expenses $ _____

    Artist   _____

                 Total $ _____

CATERING

    Estimated cost $ _____ $ _____

MUSIC

    Buy records/tapes $ _____

    Original music $ _____

                 Total $ _____

ADDITIONAL VISUALS

    Photos $ _____

    Video   _____

    Film   _____

    Other   _____

                 Total $ _____

POSTAGE/DELIVERY SERVICE

    Stamps $ _____

    Delivery $ _____

                 Total $ _____

PHOTOCOPYING

    Estimated cost $ _____

    Supplies   _____

                 Total $ _____

SET

    Planning $ _____

    Construction   _____

                 Total $ _____

Figure A–3 *(cont.)*

CAST

    Expenses $ _____

    Per diem   _____

                  Total $ _____

DUB COST

    Cost per tape $ _____

    No. cassettes   × _____

    Subtotal $ _____

    Dubbing fee $ _____

    No. cassettes   × _____

    Subtotal $ _____

                  Total $ _____

WARDROBE

    Buy $ _____

    Rent   _____

                  Total $ _____

MAKEUP/HAIR

    Supplies $ _____

    Misc. expense $ _____

                  Total $ _____

TOTAL ALL $ _____

MISC. EXPENSES

    Total all $ _____

    Add 10%   × .10 _____

                  Total $ _____

YOUR FEE $ _____

TOTAL BUDGET $ _____

Figure A–3 *(cont.)*

Show Title _____

Date _____

EQUIPMENT LIST

CAMERAS

    Total no. _____

    Format (specify number)

    VHS _____   Beta _____   8 MM _____

    Designated duties

       #1 _____

       #2 _____

       #3 _____

       #4 _____

MICROPHONES

    Total no. _____

    Type (Specify number of each)

    Handheld _____   Boom _____

    Mike stand _____   Lavaliere _____

LIGHTING

    Total no. units _____

    Type (Specify number)

    Pro lamps _____   Shiny board _____

    Camera mounted _____   Other _____

TAPE

    Total no. cassettes _____

    Kind (Specify number)

    VHS _____   Beta _____   8 MM _____

Figure A–4   Equipment List.

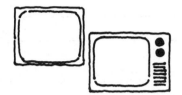

TVs/MONITORS

    (Number and size of screen)

    No. Black & White _____

    No. Color _____

ACCESSORIES

    (Specify number)

    Tripods/stands

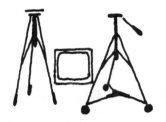

        Camera _____   Lighting _____

        Audio _____   Dollys _____

    Lighting gels (color and number of each)

        color _____   # _____

        color _____   # _____

VCRs

    Total no. _____

    Kind (specify number)

    VHS _____   Beta _____   8 MM _____

Figure A–4 *(cont.)*

Show Title _____

Date _____

CREW LIST

DIRECTOR _____

LIGHTING DIRECTOR _____

TECHNICAL DIRECTOR _____

AUDIO DIRECTOR _____

SET DESIGNER _____

SET CONSTRUCTION _____

VIDEOTAPE OPERATOR _____

CAMERA OPERATOR #1 _____

CAMERA OPERATOR #2 _____

CAMERA OPERATOR #3 _____

CAMERA OPERATOR #4 _____

GAFFER _____

GRIP _____

MAKEUP ARTIST _____

HAIR STYLIST _____

WARDROBE _____

PRODUCTION ASSISTANT _____

_____

ASSISTANT TO:

CAMERA 1 _____

CAMERA 2 _____

CAMERA 3 _____

CAMERA 4 _____

GO-FER _____

Figure A–5 Crew List.

Show Title _____

Date _____

SCENARIO

(Add to this as needed)

Location   **Sequence of Events**

_____   1. _____

_____   _____

_____   2. _____

_____   _____

_____   3. _____

_____   _____

_____   4. _____

_____   _____

_____   5. _____

_____   _____

_____   6. _____

_____   _____

_____   7. _____

_____   _____

_____   8. _____

_____   _____

_____   9. _____

_____   _____

_____ 10. _____

_____   _____

_____ 11. _____

_____   _____

Figure A–6 Scenario form.

———————————— 12. ————————————————

—————————————————————

———————————— 13. ————————————————

—————————————————————

———————————— 14. ————————————————

—————————————————————

———————————— 15. ————————————————

—————————————————————

———————————— 16. ————————————————

—————————————————————

Figure A–6 *(cont.)*

Show Title _____

Date _____

FORMAT FORM

FORMAT NO. _____ (Same as scenario no.)

    DESCRIPTION: _____

    _____

    SCENE LOCATION: _____

        INTERIORS: Yes _____   No _____

            Where? _____

        EXTERIORS: Yes _____   No _____

            Where? _____

SPECIAL

    MUSIC: Yes _____   No _____

        What? _____

    SOUND: Yes _____   No _____

        What? _____

    LIGHTING: Yes _____   No _____

        What? _____

    SET: Yes _____   No _____

        What? _____

    PROPS: Yes _____   No _____

        What? _____

    SPECIAL ELEMENTS: Yes _____   No _____

        What? _____

        _____

        _____

Figure A–7 Format form.

SHOTS TO TAPE:

Scene No.

———— 1. _____

_____

———— 2. _____

_____

———— 3. _____

_____

———— 4. _____

_____

———— 5. _____

_____

———— 6. _____

_____

TAPING DATE _____

Figure A–7 *(cont.)*

Figure A–8 Storyboard grid.

Show Title _____

Date _____

CAST LIST

ROLE _____ TALENT _____

ROLE _____ TALENT _____

ROLE _____ TALENT _____

ROLE _____ TALENT _____

ROLE _____ TALENT _____

ROLE _____ TALENT _____

ROLE _____ TALENT _____

ROLE _____ TALENT _____

ROLE _____ TALENT _____

ROLE _____ TALENT _____

ROLE _____ TALENT _____

ROLE _____ TALENT _____

EXTRAS _____

_____

_____

_____

_____

_____

_____

Figure A–9 Cast List.

Show Title _____

Date _____

Locations:
Interior   Exterior

LOCATION LIST

FIELD LOCATIONS	EXTERIORS	INTERIORS
_____	_____	_____
_____	_____	_____
_____	_____	_____
_____	_____	_____
_____	_____	_____
_____	_____	_____

STUDIO LOCATION

_____   _____   _____

Figure A–10   Location List.

Show Title _____

Date _____

PROP LIST

ITEM   NEEDED AT LOCATION

_____   _____
_____   _____
_____   _____
_____   _____
_____   _____
_____   _____
_____   _____
_____   _____

Figure A–11   Prop List.

Show Title _____

Date _____

SHOT LIST

LOCATION _____

Note: I = Interior   E = Exterior

Framing: CU, MS, WS, ECU, MCU, etc.

I or E	SHOT #	FRAMING	DESCRIPTION
___	_____	_____	_____
___	_____	_____	_____
___	_____	_____	_____
___	_____	_____	_____
___	_____	_____	_____
___	_____	_____	_____
___	_____	_____	_____
___	_____	_____	_____
___	_____	_____	_____
___	_____	_____	_____
___	_____	_____	_____
___	_____	_____	_____
___	_____	_____	_____
___	_____	_____	_____
___	_____	_____	_____
___	_____	_____	_____

CALL TIME _____

DAY _____

Figure A–12 Shot List.

CALLED CREW (if all, just say "ALL")

_____

_____

_____

_____

CALLED CAST (if all, just say "ALL")

_____

_____

_____

_____

_____

Figure A–12 *(cont.)*

SHOT LOG SHEET

Show Title _____   Producer _____

Date _____   Location _____

Page _____ of _____

Camera No. _____   Camera Operator _____   Camera Assistant _____

COUNTER NUMBER

CASSETTE #	SHOT #	TAKE #	IN	OUT	APPROX. TIME	NOTES

Figure A–13 Shot Log sheet.

Show Title _____

Date _____

EDIT LIST

CAMERA #	CASSETTE #	SHOT #	COUNTER NUMBERS	CONTENT NOTE
_____	_____	_____	_____	_____
_____	_____	_____	_____	_____
_____	_____	_____	_____	_____
_____	_____	_____	_____	_____
_____	_____	_____	_____	_____
_____	_____	_____	_____	_____
_____	_____	_____	_____	_____
_____	_____	_____	_____	_____
_____	_____	_____	_____	_____
_____	_____	_____	_____	_____
_____	_____	_____	_____	_____
_____	_____	_____	_____	_____
_____	_____	_____	_____	_____
_____	_____	_____	_____	_____
_____	_____	_____	_____	_____

MATERIALS TO TAKE TO THE EDIT

Special art: _____

_____

_____

_____

End credits: _____

Title cards: _____

Figure A–14 Edit List.

Show Title _____

Date _____

20-MINUTE PREPRODUCTION PLAN

TIME ALLOTTED	YOUR TIME	
00:30	_____	WHAT DO YOU WANT TO DO? (Briefly)
		_____
03:30	_____	WHERE WILL YOU TAPE? List the locations then on a separate piece of paper, draw a basic sketch of each one.

LOCATION

#1 _____

#2 _____

#3 _____

#4 _____

| 01:00 | _____ | WHO IS IN THE SHOW? List your cast as you now know it: principals, supporting, voice over, extras. |

ROLE   TALENT

_____   _____

_____   _____

_____   _____

_____   _____

_____   _____

_____   _____

Figure A–15 20-Minute Preproduction Plan.

01:00 _____ WHAT IS YOUR EQUIPMENT AND CREW?
List the equipment and the crew assigned to it:
cameras, production assistants, grips, gaffers, go-fers,
and so on.

EQUIPMENT CREW

_____ _____

_____ _____

_____ _____

_____ _____

_____ _____

02:00 _____ WHERE WILL YOU PUT YOUR CAMERAS?
Using the sketches you drew for each location,
indicate exactly where you will put your cameras.

01:00 _____ WHERE WILL ANY REMOTE MICROPHONES BE
PLACED? Using the location sketches, indicate the
position of any microphones. Also list them here:

_____

_____

_____

01:00 _____ WHERE IS THE LIGHT AT THE LOCATIONS? Add to
location sketches the position of any lighting. Add to
this any lighting that you propose to take in. Also list
this additional lighting equipment here:

_____

_____

_____

Figure A–15 *(cont.)*

03:00   _____   **WHAT WILL THE SHOW LOOK LIKE?** Using the form provided here, draw storyboards for the three main shots of the video. Make these simple but complete enough to convey the idea.

0.6:00   _____   **WHAT IS THE SHOT LIST?** List every shot you can think of that you will want. Note the framing of these shots, such as CU, WS, MS, ECU, and MW.

SHOT #   FRAMING   DESCRIPTION

_____   _____   _____

_____   _____   _____

_____   _____   _____

_____   _____   _____

_____   _____   _____

_____   _____   _____

_____   _____   _____

_____   _____   _____

_____   _____   _____

_____   _____   _____

Figure A–15   (*cont.*)

01:00   _____   WHAT IS THE SCHEDULE? When and where will you
                                  tape? Day? Time? Place? And who is to report on
                                  those taping days?

PLACE _____

DAY _____

TIME _____

CREW CALLED _____

_____

_____

_____

CAST CALLED _____

_____

_____

_____

_____

_____

_____

Your plan is now complete. ROLL TAPE!

Figure A–15   *(cont.)*

Show Title _____

Date _____

15-MINUTE POSTPRODUCTION PLAN

TIME   YOUR
ALLOTTED   TIME

02:30   _____   ARE THERE OTHER THINGS TO SHOOT? Check your
                           shot list to make sure you have recorded all shots.
                           List any shots you still need, including graphics,
                           photos, cutaways, and the like.

                           SHOT #   FRAMING   DESCRIPTION
                           _____ _____ _____
                           _____ _____ _____
                           _____ _____ _____
                           _____ _____ _____
                           _____ _____ _____

01:00   _____   ARE THERE ANY TITLES OR CREDITS? Write down the
                           name of the show and any credit you want to include.
                           Do not write in specific names for cast or crew; you
                           have these already listed on your 20-minute
                           preproduction plan. Note only that you want to
                           include cast and crew in the credits, if you do.

                           TITLE _____

                           CREDITS: Cast _____   Crew _____

                           Others _____
                           _____
                           _____

01:00   _____   DO YOU WANT TO ADD ANY ADDITIONAL AUDIO?
                           Music? Announcer?

                           _____
                           _____
                           _____
                           _____
                           _____
                           _____

Figure A–16   15-Minute Postproduction Plan.

02:00 _____ WHAT EFFECTS DO YOU WANT TO ADD IN THE EDIT?
Are there any special effects, audio or video, that you
want to add?

AUDIO EFFECTS

DESCRIBE _____

_____

_____

VIDEO EFFECTS

DESCRIBE _____

_____

_____

08:30 _____ MAKE THE EDIT LIST. Using the shot list, make an
edit list in the exact order that you want the shots cut
together in the final show.

CAMERA	CASSETTE	SHOT	COUNTER NUMBER
_____	_____	____	_____
_____	_____	____	_____
_____	_____	____	_____
_____	_____	____	_____
_____	_____	____	_____
_____	_____	____	_____
_____	_____	____	_____
_____	_____	____	_____
_____	_____	____	_____
_____	_____	____	_____
_____	_____	____	_____
_____	_____	____	_____
_____	_____	____	_____
_____	_____	____	_____
_____	_____	____	_____
_____	_____	____	_____

You are now ready to edit.

Figure A–16 (*cont.*)

INDEX